[redacted]

AF182335

Petroleum Resources

with Emphasis on Offshore Fields

WITPRESS

WIT Press publishes leading books in Science and Technology.
Visit our website for the current list of titles.
www.witpress.com

WITeLibrary

Home of the Transactions of the Wessex Institute, the WIT electronic-library provides the international scientific community with immediate and permanent access to individual papers presented at WIT conferences. Visit the WIT eLibrary at http://library.witpress.com

REF
TN
871.3
.G75
2010

Petroleum Resources

with Emphasis on Offshore Fields

O.T. Gudmestad
University of Stavanger, Norway

A.B. Zolotukhin
Gubkin Russian State University of Oil and Gas, Moscow, Russia

E.T. Jarlsby
PETRAD, Stavanger, Norway

BRAZOSPORT COLLEGE LIBRARY

WIT*PRESS* Southampton, Boston

NOV 10 2011

O.T. Gudmestad
University of Stavanger, Norway

A.B. Zolotukhin
Gubkin Russian State University of Oil and Gas, Moscow, Russia

E.T. Jarlsby
PETRAD, Stavanger, Norway

Published by

WIT Press
Ashurst Lodge, Ashurst, Southampton, SO40 7AA, UK
Tel: 44 (0) 238 029 3223; Fax: 44 (0) 238 029 2853
E-Mail: witpress@witpress.com
http://www.witpress.com

For USA, Canada and Mexico

WIT Press
25 Bridge Street, Billerica, MA 01821, USA
Tel: 978 667 5841; Fax: 978 667 7582
E-Mail: infousa@witpress.com
http://www.witpress.com

British Library Cataloguing-in-Publication Data

A Catalogue record for this book is available
from the British Library

ISBN: 978-1-84564-478-9

Library of Congress Catalog Card Number: 2009934443

The texts of the papers in this volume were set individually by the authors or under their supervision.

No responsibility is assumed by the Publisher, the Editors and Authors for any injury and/or damage to persons or property as a matter of products liability, negligence or otherwise, or from any use or operation of any methods, products, instructions or ideas contained in the material herein. The Publisher does not necessarily endorse the ideas held, or views expressed by the Editors or Authors of the material contained in its publications.

© WIT Press 2010

Printed in Great Britain by MPG Books Group, Bodmin and King's Lynn.

All rights reserved. No part of this publication may be reproduced, stored in a retrieval system, or transmitted in any form or by any means, electronic, mechanical, photocopying, recording, or otherwise, without the prior written permission of the Publisher.

Biographies

Professor Ove Tobias Gudmestad has been adjunct professor at the University of Stavanger since 1994 and, from 2008, full professor of Marine Technology at the University of Stavanger, Norway.

Gudmestad has experience from engineering, research and projects in Statoil from 1975 to 2008. His offshore-related experience includes pipeline design, offshore structures design, arctic engineering and field development studies. He has management experience from several research projects, concept development studies and engineering tasks.

Gudmestad has published a number of papers and filed several patent applications. He gained an MSc in applied mathematics from the University of Tromsø, Norway in 1973 and a PhD in hydrodynamics from the University of Bergen in 1985. For two years he studied at Massachusetts Institute of Technology, USA.

Since 2005, he has also been adjunct professor in arctic offshore civil engineering at The Norwegian University of Science and Technology in Trondheim. The Gubkin Russian State University of Oil and Gas in Moscow and the Murmansk State Technical University awarded him honorary doctorate degrees in 2002 and 2008 respectively. He is a Member of the Norwegian Academy of Technological Sciences and a Fellow of Royal Institution of Naval Architects, UK.

Professor Anatoly Zolotukhin holds the position of Deputy Rector on International Affairs of the Russian State Gubkin University of Oil and Gas (Moscow, Russia). He holds a MS degree in petroleum engineering (1969), MS degree in mathematics (1976), PhD in fluid mechanics (1973) and a Doctor of Science degree in reservoir engineering (1990). He has more than 30 years of academic and working experience in areas of reservoir and production engineering, EOR/IOR methods, multiphase flow of fluids in porous media and application of different approaches to reservoir characterization, evaluation of reserves, production forecast and field development modelling. Anatoly B. Zolotukhin is an author and co-author of 12 books and more than 100 articles.

From 1999 till 2004 Anatoly Zolotukhin worked in Statoil as a project manager, and from 2004 until 2008 – as a Technical Director of Statoil representative office in Moscow, Russia.

Anatoly Zolotukhin is a member of the Russian Academy of Natural Sciences.

In addition to his main job he holds several other positions:

- Professor International at Stavanger University (Stavanger, Norway)
- Doctor Honoris Causa, Murmansk State Technical University (Murmansk, Russia)
- Doctor Honoris Causa, Arkhangelsk State Technical University (Murmansk, Russia)
- Vice-President of the World Petroleum Council (since 2008)
- Member of the Academic Council of the European Energy Forum (EUREF) Institute (since 2008)

Dr Erik T. Jarlsby holds a Lic.oec. degree in international economics from the University of St.Gallen, Switzerland, and a PhD in internatonal business from the University of Twente, Netherlands, with as doctoral dissertation on hydrocarbon markets. He has worked for Mobil, Statoil and Borealis, holding professional and managerial positions in field development analysis, NGLs trading and petrochemical feedstock supply. He has lectured in numerous countries on business strategy, petroleum downstream & natural gas issues and project economic analysis. Since 2002, Dr Jarlsby works as a consultant and lecturer through his company Erik Jarlsby AS based in Stavanger, Norway. He is a regular contributor to the PETRAD courses on petroleum development and resource management.

Table of Contents

Preface

This book on *Petroleum Resources with Emphasis on Offshore Fields* addresses the challenges of transforming hydrocarbons that exist in the underground, to valuable products that can be sold and delivered. It is intended for readers who have a professional or student interest in the petroleum industry, and a basic level of prior knowledge in the technical and commercial aspects of the industry. The goal of the book is to increase the reader's general understanding of key work processes in the "upstream" part of the petroleum industry, that is, the part of the industry that locates underground resources and converts them to valuable products.

The need for a book on the theme of *Petroleum Resources with Emphasis on Offshore Fields* arises from several conditions:

- Much work in the petroleum industry is carried out in *interdisciplinary project teams* and work processes. Involved personnel need to understand the wider context of petroleum resource development, and to share a terminology related to this.
- Decision makers need to *evaluate all aspects of the value chain* to identify optimal field development solutions.
- There is a need to recruit *new graduates* to the industry to work for oil companies and the engineering and service industry. Recruits to the industry will benefit from a general introduction to key topics related to developments of petroleum resources.
- Courses in the field of petroleum developments need to be supported by a *reference book* that can supplement presentations and serve as a learning resource for students.
- Scientists in different areas of the offshore oil and gas industry lack in many aspects the full understanding of the complexity of the industry. There is thus the need for a book that will serve as a reference book for scientists who need to increase their knowledge about other aspects than their own area of expertise in order to carry out their research or become involved in interdisciplinary research projects.

This book was initially written as reference book for training programs, notably Petrad's 8-week course on "Petroleum Developments and Operations," and the course on "Offshore Field Development" at the University of Stavanger. These courses are attended by students from many parts of the world, including areas that are in early stages of hydrocarbon developments as well as Norway with its established offshore petroleum provinces. Although intended to be useful in a wide range of contexts world wide, it has particular emphasis on offshore developments, and is based to a large extent on insights from Norwegian offshore projects.

The scope of the book has thereafter been extended to incorporate the needs of scientists that need an interdisciplinary view of the oil and gas industry to carry out their research ambitions.

The cooperation with a number of former colleagues in Statoil and with the management of the Petrad Foundation is very much appreciated. Particular assistance has been received from Bernt Thorbjørnsen, Oddbjørn Tolo, and Jonas Odland from Statoil and Ellinor Melbye from Petrad. Alfred Kjemperud and Bjørn-Erik Leerberg have provided valuable background on licensing systems. Kjell Traa has contributed as co-author of parts of Chapters 4–6 and 10.

Acknowledgments are also extended to John W.A. Coker, who first led a course on "Management of Petroleum Development and Operations" for Petrad.

A group of enthusiastic students assisted to prepare the first draft manuscript of the book: Anna Elofsson, Anja Holmefjord, Frida Mathayo, Bård Solvang, and Mathilde Wilhelmsen. Their efforts are greatly appreciated. Thanks are also extended to Akaradech Apornsuvan who has assisted to prepare figures for the book.

It is the hope that the book will be of value to the readers and that it will serve its purpose as reference book for those, including the researchers, needing an interdisciplinary reader on the subjects involved in the development of petroleum resources.

The Petroleum Safety Authority (PSA) of Norway (www.ptil.no), located in Stavanger, has supported the publication of the book, while Aker Exploration, Norway, and Aker Geo, Norway, have contributed with funds to enable printing of the book. Their support is greatly appreciated.

Ove T. Gudmestad, Anatoly B. Zolotukhin, and Erik T. Jarlsby

University of Stavanger, Stavanger and
Gubkin State University of Oil and Gas, Moscow, 2010

Introduction

There are large challenges facing the oil and gas industry, for example, increasing challenges to find new oil and gas accumulations, a requirement for a clean environment and reductions in CO_2 emissions being voiced by public and governments, requirement for long-distance transport of natural gas from the producers to the consumers, and rapid changes in the price of hydrocarbons in the world market. The decision makers need a wide background for their decisions, and all aspects of the value chain are important in order to identify the optimal field development solution.

The objective of this book is therefore to treat in a consistent manner the key aspects related to development of petroleum resources in order to allow an *interdisciplinary view* of the topics involved in development projects and in order to *identify the most optimal development solution.* The book is organized to allow the reader to review the different topics individually so that the book can serve as a reference source for those interested in topics not familiar to themselves.

Chapters 1–6 follow the hydrocarbons in a production system from their origins in reservoirs through reservoir management, wells, and flow assurance to processing and offtake. Geology, reservoir, and production are the key elements in the development of petroleum resources. The drilling, well design, and well completion aspects are complementary to the subsurface aspects to ensure a profitable project. When the wellstream has reached the wellhead, it must be processed to meet the requirements of products that can be sold. For transport of hydrocarbons over longer distances, the flow in pipelines has to be assured, and long-distance flow of non-treated wellstream is discussed in the book.

Facilities for oil and gas production including subsea as well as surface facilities are reviewed in Chapter 7. The process of planning a project is reviewed in Chapter 8 with an emphasis on the early project stages. Furthermore, field decommissioning is reviewed, as the restoration of the environment is considered an important responsibility for the industry (Chapter 9). Health, safety, and environmental issues are central to any modern petroleum development, and are addressed in Chapters 10 and 11.

Chapter 12 on licensing and fiscal regimes reviews the ways in which governments may grant rights to explore for and produce petroleum to firms, and the sharing of the economic results between the state and firms. Chapter 13 on economics reviews the financial structure of petroleum firms and methods for analyzing the profitability of investment projects, with supporting information on cost estimation and analysis of uncertainty.

The petroleum industry handles very large amounts of financial value. There are cases in the world where the industry has mainly benefited a small section of society, while many others have seen no significant benefit or even a deterioration of their living conditions as a result of the way the industry has been managed. Ethical issues and the responsibilities of the industry to society are therefore discussed in the final chapter.

The geology of petroleum resources

Geology is the study of the earth – its history, structure, composition, life forms, and the processes that continue to change it. Hydrocarbons, among other resources, are found within the earth's subsurface.

The development of any petroleum field must be based on insights into the geology of the underground formations from which the oil and gas will be extracted. The disciplines of geology and geophysics provide information on the location and estimated volumes of petroleum in the underground. Based on such information, the reservoir engineer estimates how much recoverable resources one can expect based on the current technology. Such estimates are associated with a series of uncertainties, as is inevitable when dealing with subsurface resources.

This chapter is intended as an introduction to geology, with relevance for petroleum developments. The chapter introduces petroleum geology, basin analysis, prospect evaluation, and volumetric estimation and uncertainty associated with the estimation.

1.1 The formation of hydrocarbon resources

Petroleum resources are the result of formation processes spanning millions of years. For the process to be completed, several key elements must be present, each of which will be reviewed further below:

- *Source rock*: Rock containing organic matter that can evolve into hydrocarbons.
- *Maturation*: The process of converting organic matter to hydrocarbons.
- *Migration*: The movement of hydrocarbons from source into a reservoir.
- *Reservoir*: Rock that can contain hydrocarbons due to its porosity and permeability.
- *Trap*: A configuration of reservoir rock in which hydrocarbons can accumulate.
- *Seal*: Impermeable rock preventing hydrocarbons from escaping from the reservoir.
- *Timing*: A sequence of forming the above elements so as to allow hydrocarbons to accumulate.

1.1.1 Source rock

A source rock is a rock rich in organic matter, which, if heated sufficiently, will generate oil or gas. All living things are part of the carbon cycle: plants use photosynthesis to convert carbon dioxide in the atmosphere to biomass, and carbon dioxide is returned to the atmosphere as dead plants decay or by breathing animals. From this cycle, a tiny portion (0.1%) of carbon will be trapped as the forerunner to fossil fuels, that is, coal, oil, or gas. The prerequisite for this entrapment is that debris of organisms is deposited in oxygen-depleted water, usually in deep stagnant waters. Continued sedimentation causes further burial of the debris and hence compaction.

Under the combined effect of temperature, pressure, and time, the organic deposits eventually transform into kerogen and thereafter oil or gas. Kerogen is a substance formed from organic matter which, under certain conditions, can generate oil or gas. It is common to distinguish four types of kerogen. They differ in their origins (marine or land material, etc.), their ratios of hydrogen to carbon, and their tendency to generate oil, gas, or coal. Rocks of marine origin are oil prone, while those of terrestrial origin tend to be gas prone.

Source rock evaluations are done by methods of geochemistry to identify the nature of organic-rich rocks, which contain the precursors to hydrocarbons. Using geochemistry, the type and quality of the generated hydrocarbons can be assessed.

1.1.2 Maturation

Over time and depending on temperature, kerogen will undergo maturation and generate hydrocarbons (catagenesis). There is a considered range of depth (see Figure 1.1) and temperature interval, the "oil window," required for oil to mature in the source rock. On average, the oil window varies from 60 °C to 175 °C with the peak at 90 °C (see Figure 1.2). Gas can be produced above and below the oil window.

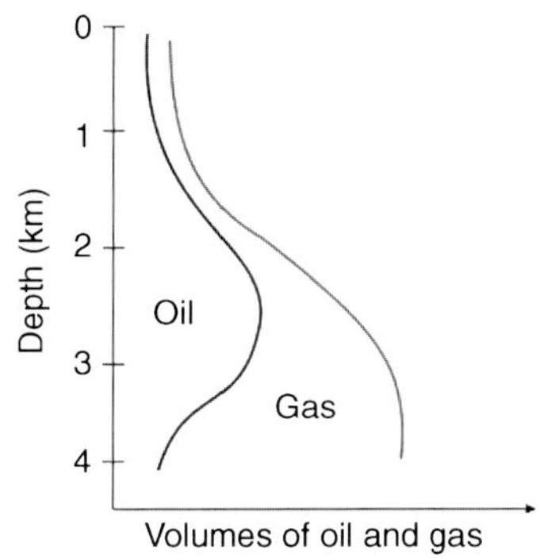

Figure 1.1: "Oil window," depth to the source rock where oil may be generated.

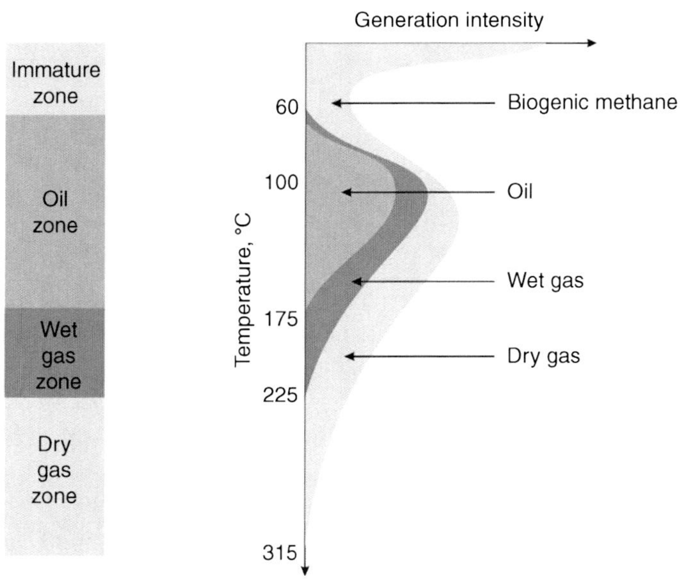

Figure 1.2: Oil window of the source rock with regard to temperature.

The time in which the source rock is exposed to temperature is vital for maturation. While temperature has an exponential influence factor, time has a linear influence relationship. This implies that younger rocks will require high temperature. Doubling the time doubles the maturation.

Depending on the geothermal gradient, the depth of the oil window varies from one region to another. As an example, the North Sea has a geothermal gradient of 30 °C/1000 m, and its oil window ranges from 1500 m to 6000 m.

Maturation is assessed by studying the history of the source rock in order to obtain predictions of the amount and timing of hydrocarbon generation and expulsion.

1.1.3 Phase diagram and chances of finding petroleum

A phase diagram is a graphical description of different fluid phases that can be found in different ranges of temperature, depth, and pressure. It gives an overview of what can be found at certain depths under normal circumstances.

Figure 1.3 is a phase diagram. Point C is called a *critical point.* It is the temperature and pressure where the vapor and liquid phases coexist, and the properties of the vapor phase and the liquid phase are the same. This is an experimentally determinable point. As shown in the diagram, there is a considerable region at temperatures higher than the critical point that is two phases, and a smaller region of pressures higher than the critical pressure that is also two phases. The highest temperature, at which two phases can coexist, point C_T, is called *cricondentherm.* The highest pressure, at which two phases can coexist, point C_P, is called *cricondenbar.*

1.1.4 Migration

Migration refers to the movement of hydrocarbons from their source to reservoir rocks. The movement of newly generated hydrocarbons out of their source is the *primary migration,* also called expulsion. The further movement of the hydrocarbons into reservoir rock in a hydrocarbon

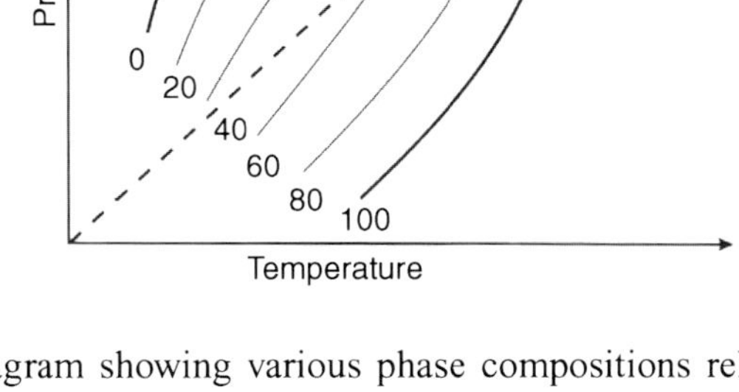

Figure 1.3: A phase diagram showing various phase compositions relative to depth, pressure, and temperature (Adapted from Zolotukhin *et al.*, 2000).

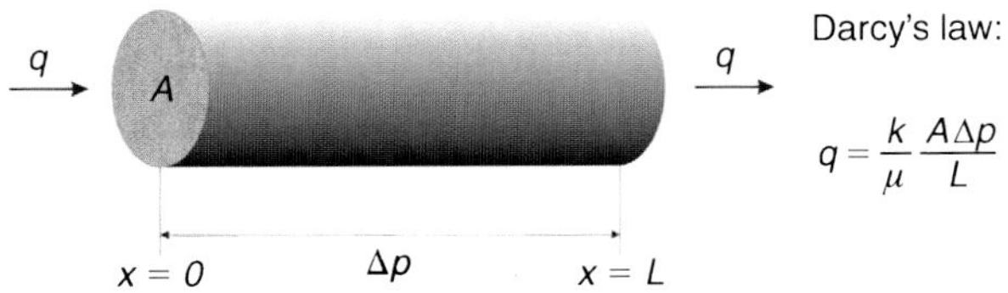

Figure 1.4: A schematic illustration of the fluid flow in a porous rock, where k is the permeability, μ the fluid viscosity, Δp the pressure difference across the pore boundary, L the length, q the volumetric flow rate, A the cross-sectional area, and (q/A) the superficial (Darcy) velocity (volumetric flow rate divided by the cross-sectional area normal to flow).

trap or other area of accumulation is *secondary migration*. Migration typically occurs from a structurally low area to a higher area because of the relative buoyancy of hydrocarbons in comparison to the surrounding rocks. Migration can be local or can occur along distances of hundreds of kilometers in large sedimentary basins, and is critical to the formation of a viable petroleum system.

The host rocks are normally saturated with water before the arrival of any migrating hydrocarbons. The buoyancy effect will result in gas moving higher compared to oil and water. Simultaneously, migration can be vertical and/or lateral in the direction of lower pore pressure. The study of migration reveals information on hydrocarbons' move from source to reservoir, and helps to identify the sources of hydrocarbons in the particular area.

1.1.5 Reservoir rocks

A reservoir is a subsurface body of rock having sufficient porosity and permeability to store and transmit fluids. Analysis of a reservoir requires information on the porosity and permeability of a rock.

Porosity is a measure of pore spaces in a rock and is expressed in percentage of the total volume. The estimate of the porosity is useful in quantifying the volume of hydrocarbons in place. Porosity is classified as primary or secondary depending on its origin.

The permeability of a rock is a measure of the ability of liquid to flow through it. Permeability is measured in Darcy, D, or milli-Darcy, mD. For a rock to be permeable, the pores must be connected. The flow of liquids through rock depends on the rock permeability, the fluid viscosity, and the pressure variance across any pore boundary. Oil reservoirs which can be developed profitably typically have a porosity of 10–20% and a permeability of 50–2000 mD. Figure 1.4 illustrates the permeability concept.

1.1.6 Traps

A trap is a configuration of rocks suitable for containing hydrocarbons. It must be sealed by a relatively impermeable formation through which hydrocarbons will not migrate. It is simply a place where oil accumulates and is barred from further escape. If there are enough hydrocarbons, traps will be filled to spill point and will move to the next trap or spill out to the surface.

Traps are classified as *structural* if they formed as a result of strata deformation by tectonic movements, or *stratigraphic* if they are a result of generic changes in the sediment itself. Normally structural traps are easier to find than stratigraphic traps.

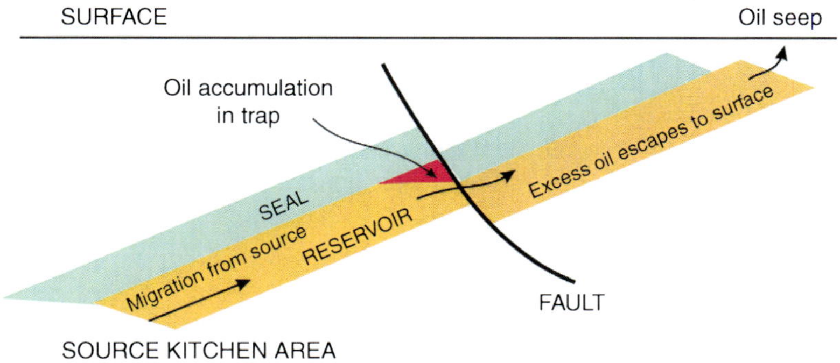

Figure 1.5: A schematic illustration of the key elements for petroleum accumulation as discussed in the text. (Adapted from an illustration by C. Ferrero).

1.1.7 Seals and cap rock

The seal or cap rock is a unit with low permeability that impedes the escape of hydrocarbons from the reservoir rock. Good seals should be impermeable and ductile, forming a barrier above and around the reservoir rock. Thickness and extension are other aspects to be considered in analyzing the cap rock. Figure 1.5 shows the classical setting of different rocks as mentioned above.

For hydrocarbon accumulations to develop in traps, the above-mentioned elements must be formed in an appropriate sequence. Traps, for example, need to form before oil migrates from the source.

1.2 Basin analysis

Oil and gas are located only in certain limited locations below ground. Finding these locations, thousands of meters below surface, is obviously quite a challenge. It is also an expensive and commercially risky undertaking: drilling an exploration well may cost several million dollars onshore, and several tens of million dollars offshore, having often a 90% probability of finding nothing of value.

Petroleum geologists have therefore developed methods for analyzing the geology to make the best possible use of available information as guidance for further exploration activity. This entails describing and quantifying the physical processes by which sedimentary basins develop, and how these processes operate over a wide range of spatial and temporal scales. This section will review some important concepts and terminology which are applied in this analysis.

The information required for analysis can be obtained through geological, petrophysical, and geophysical interpretations, including the study of outcrops, cores, logs, rock cuttings, and seismic data.

1.2.1 Basin

A basin is a depression in the crust of the earth, caused by plate tectonic activity and subsidence, in which sediments accumulate. A basin usually refers to a large area (several thousand square kilometers). Sedimentary basins vary from bowl-shaped to elongated troughs. Basins can be bounded by faults. If rich hydrocarbon source rocks occur in combination with appropriate depth and duration of burial, then a petroleum system can develop within the basin.

1.2.2 Petroleum system

A petroleum system consists of a mature source rock, reservoir, and a seal. Appropriate relative timing of formation of these elements and the processes of generation, migration, and accumulation are necessary for hydrocarbons to accumulate and be preserved.

Exploration plays and prospects are typically developed in basins or regions in which a complete petroleum system has some likelihood of existing.

1.2.3 Play

A play is defined as a group of prospects within a geographically delimited area, where a set of mutually related geological factors must be present concurrently in order to permit the discovery of hydrocarbons. A basin can have one or more plays. A play provides an explanation of how petroleum may have been formed in the area, and is also referred to as an exploration model.

The geological factors are typically reservoir rocks, traps, mature source rocks, and migration paths, plus the condition that the traps were formed before the migration of hydrocarbons ceased. All fields, discoveries, and prospects within the same play are characterized by the play's specific set of geological factors and can therefore be distinguished from the fields, discoveries, and prospects belonging to other plays.

1.2.4 Prospect

A prospect is a trap with reservoir rock in which hydrocarbons have been predicted to exist. A prospect is commonly an anomaly, such as a geological structure or a seismic amplitude anomaly,

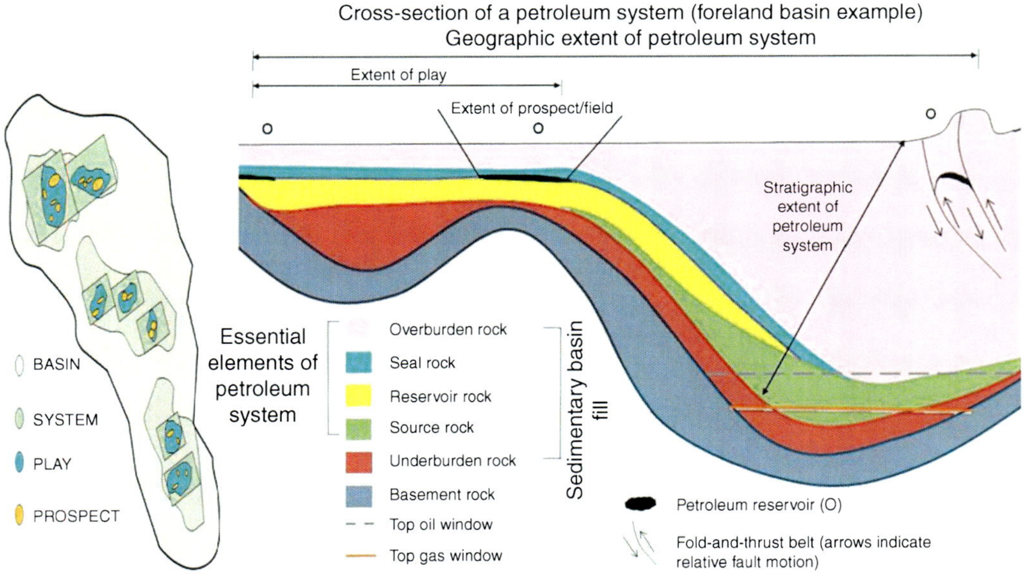

Figure 1.6: Schematic illustration of the relationship between petroleum system, play, and prospect/field. The left figure is a plan view and the right figure is a cross-section view. (Adapted from Magoon & Dow, 1994).

that is recommended by explorationists for drilling a well. A trap and the timing of generation and migration are the key factors for an existence of a prospect. A group of different reservoir rocks stacked upon each other within a trap is also called a prospect. A group of prospects of similar nature constitutes a play. Only prospects that have been evaluated to have the potential of containing economic quantities of hydrocarbon are investigated in further details.

1.2.5 Discovery (field)

While seismic surveys and interpretation can reveal prospects with some probability of containing hydrocarbons, it is only by drilling a well that the hydrocarbons can really be found. A discovery, or hydrocarbon field, is a prospect that has been confirmed by drilling to contain hydrocarbons. Prior to drilling a well, the probability of finding hydrocarbons in a prospect is usually low, often around 10%.

A hydrocarbon field consists of a reservoir in a shape that will trap hydrocarbons (see Figure 1.6) and that is covered by an impermeable, sealing rock. Typically, the term implies that the quantity of hydrocarbons present is sufficient to enable a profitable development.

1.3 Play and prospect analysis

Prospect evaluation involves delineation of critical rock units and making detailed interpretations of structural features prior to drilling. Since the drilling of an exploration well is an expensive activity which, in many cases, results only in a dry well, prospect evaluation prior to drilling is important in order to select as correctly as possible the locations that should be drilled. This section gives an introduction to concepts which petroleum geologists may apply to analyzing plays and prospects.

1.3.1 Play elements mapping

A play is a collection of prospects having a relatively homogeneous geological setting. Therefore, prospects in a play share some elements of common risk. These risks can be analyzed by using the available geological information in combination with statistical concepts of probability. Such analysis of play elements can provide guidance for evaluating individual prospects within the play.

A play can be analyzed in terms of the following elements:

- *Reservoir*: Reservoir rocks are mapped to identify their presence and quality.
- *Source*: A source rock has to be present with good quality and should be matured. Migration and timing connect the source rock with reservoirs.
- *Seal*: A seal having a capacity to hold hydrocarbons in reservoirs is identified.

The availability of all three elements is associated with uncertainty, and therefore each has a risk tag.

1.3.2 Play and prospect probabilities

Based on information that is available on a play, it may be possible to estimate probabilities for each of the plays. Probabilities range from 0.0 (certainty of non-occurrence) to 1.0 (certainty of occurrence), with values in between indicating different probabilities that the element is in place.

The probabilities indicate the likelihood of being able to make a discovery, but do not indicate the size of any such discovery.

Play probabilities therefore include:

- $P_{Reservoir}$: The probability of the occurrence of rock that can form reservoirs on a regional scale.
- P_{Source}: The probability of the occurrence of a rock unit that can generate and expel oil or gas in sufficient quantity to form one or more accumulations within the play.
- P_{Seal}: The probability of the occurrence of a regional top seal, capable of preventing hydrocarbons from upward migration.

Prospect probabilities include:

- $P_{Reservoir'}$: The probability of the occurrence of reservoir rock in the prospect with effective porosity and permeability, capable of holding hydrocarbons above a specified minimum volume in the prospect (assuming that such reservoir rock exists on a regional scale).
- P_{Trap}: The probability of the occurrence of a structural or stratigraphic configuration that provides a trap for migrating hydrocarbons.
- $P_{Charge/migration}$: The probability of the presence of quality and maturity of source rocks in the drainage area of the prospect, sufficient migration of hydrocarbons into the trap and in-reservoir biodegradation.

Before any prospect in a play is drilled, the probability of making a discovery in a particular prospect can be calculated as the product of all individual element probabilities:

$$P_{Discovery} = P_{Reservoir} \times P_{Source} \times P_{Seal} \times P_{Reservoir'} \times P_{Trap} \times P_{Charge/migration}$$

If a discovery is made in a prospect, then all probabilities can be set to 1.0 for the play and for that particular prospect.

1.4 Resources: terminology and estimation

After having dealt with the question of whether petroleum is present or not, there remains the question of how large the resources might be and furthermore, whether they can be developed profitably. *Resources* and *reserves* are the common terminology used to qualify the status of the subsurface hydrocarbons.

It is relevant to make estimates of potential resource sizes prior to drilling exploration wells, in order to justify the cost of such wells, but in particular such estimates are needed when a discovery has been made, and a possible development project is being contemplated. Oil and gas companies also estimate the reserves to which they have access as part of their financial reporting, and such estimates can have a big influence on how investors will judge the company, that is, influence on the company's share price, in the case of a company whose shares are traded on a stock exchange.

1.4.1 Oil and gas in place

STOOIP, or *stock tank oil originally in place*, is a term referring to the total amount of oil present in a field before production has started. Stock tank oil initially in place (STOIIP) and oil initially in place (OIIP) are other terms meaning the same thing.

For gas, the term gas initially in place (GIIP) is sometimes used, and correspondingly CIIP for condensate, but one might just as well refer to gas in place or condensate in place.

1.4.2 Resources

Resources is a collective term that covers technically recoverable quantities of petroleum. The resources are classified according to the maturity of the project that is necessary to recover them. The main classifications are reserves, contingent resources, and undiscovered resources.

The term *oil reserve* refers to the portion of oil in place (STOOIP) that is recoverable under economic constraints. Oil in the ground is not a reserve unless it is economically recoverable. The *recovery factor* (RF) is the percentage of the oil in place which is economically recoverable under a given set of conditions.

1.4.3 Methods of estimating reserves and their uncertainties

> If a man will begin with certainties he shall end in doubts, but if he will be content to begin with doubts, he shall end in certainties.
>
> Francis Bacon (1606)

Different methods of estimating resources and reserves can be divided into three groups depending on the uncertainty handling used by each method. These three groups comprise the methods based on *deterministic*, *probabilistic*, and *fuzzy/interval* calculations.

Below are general formulae for volumetric calculation of reserves:

$$\text{Reserve} = \text{STOOIP} \cdot \text{RF}$$
$$\text{STOOIP} = \frac{V_b \times \text{NTG} \times \phi \times S_{oi}}{B_{oi}}$$

where RF is the recovery factor, V_b the rock volume, NTG the net-to-gross factor, ϕ the porosity, S_{oi} the oil saturation, and B_{oi} the oil formation volume factor.

The selection of the method for estimating resources and reserves depends on the quantity and quality of the available data. Different methods of handling the involved uncertainties result in different estimates (Zolotukhin *et al.*, 2000).

1.4.4 Deterministic calculation

A numeric analysis can be *deterministic* or *probabilistic* (*stochastic*) depending on whether uncertainty has been incorporated or not. Deterministic analysis does not incorporate uncertainty, whereas probabilistic analysis does.

A deterministic method can be used when all the variables required for reserve calculation can be determined using the available technology. For example, if we have a porosity of 20%, rock volume of 23 km^3 and so forth, the result will be a single best estimate determination of the volume initially in place. Accuracy of the value assigned to each variable is subject to accuracy of interpreter, data acquisition and processing, tools used, etc. A deterministic model and its results are usually interpreted as the most likely outcome. The uncertainty in the results can be illustrated by sensitivity cases where the most important of the uncertain input parameters are changed in the model. Usually only one parameter is changed for each sensitivity case.

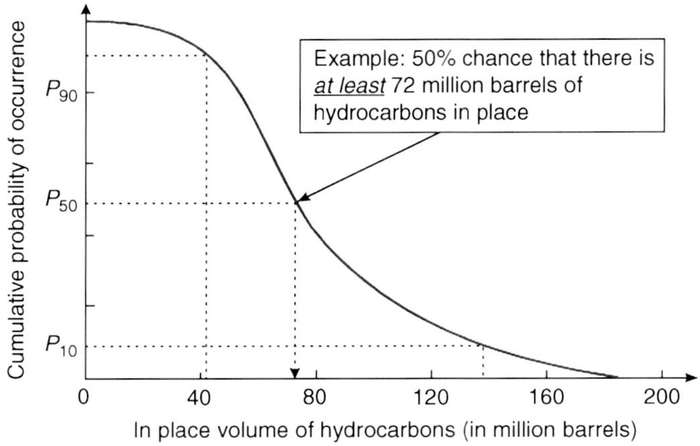

Figure 1.7: Probabilistic assessment of hydrocarbon reserves.

1.4.5 Probabilistic evaluation

The method of estimation is called probabilistic or stochastic when the known geological, engineering, and economic data are used to generate a range of estimates and their associated probabilities. The resulting reserve volume is expressed as a range of values and their associated probabilities. Probabilistic methods deal better with uncertainty but depend on mathematical assumptions and distributions that are prone to errors. Figure 1.7 illustrates an example of a probabilistic analysis result.

1.4.6 Fuzzy and interval estimation

Fuzzy and interval calculi are used when only limited information is available. They are similar in the sense that both are based on the uncertainty spread (interval). The difference is that a fuzzy number usually contains a most likely value while an interval number constitutes of an interval (minimum and maximum values). Chances of getting any number within the interval are the same (e.g., uniformly distributed within the interval).

All methods are similar in the sense that they are used for volumetric calculations and they all assess uncertainty of the input data. On the other hand, they differ in how they involve uncertainties. Figure 1.8 compares and contrasts the resultant volumes as can be calculated from different methods.

1.4.7 Example of the reserve estimate

Consider a simple example: reserves of a discovered field will be estimated by using different approaches as discussed above. Input data available for the reserve estimate are given in Table 1.1.

Results of the reserve estimate performed by using deterministic, probabilistic, fuzzy, and interval techniques are shown in Table 1.2 and Figure 1.9.

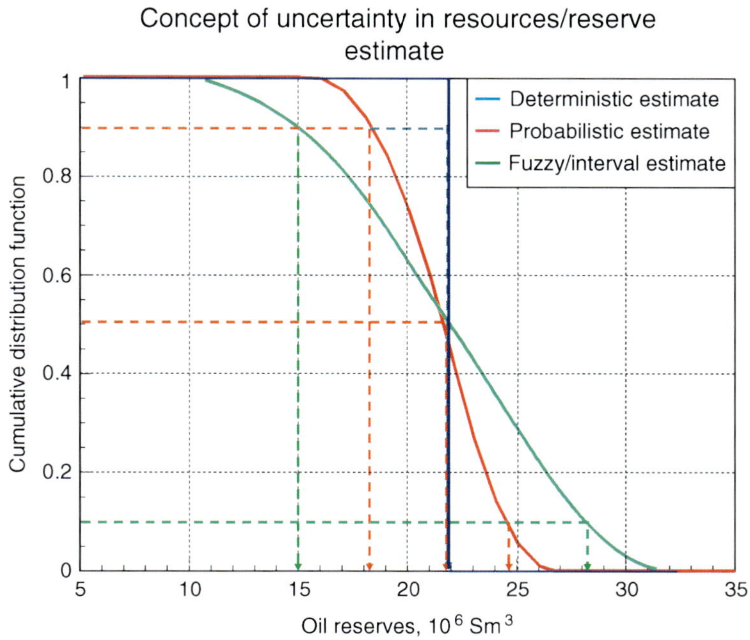

Figure 1.8: Illustration of approaches used for the reserve estimate.

Table 1.1: Input data for the reserve estimate.

Value	Gross rock volume, GRV (10^9 m^3)	Net-to-gross ratio, N/G (m/m)	Porosity, ϕ (fraction)	Initial oil saturation, S_0 (fraction)	Oil formation volume factor, FVF (Rm^3/Sm^3)	Ultimate oil recovery, OR (fraction)
Minimum	15	0.13	0.24	0.55	1.1	0.35
Most likely	77	0.23	0.28	0.65	1.15	0.4
Maximum	350	0.33	0.32	0.79	1.2	0.45

Results indicate that the spread of the results (e.g., the distance between P_{90} and P_{10}) is increasing from the deterministic estimate (zero spread) to the interval estimate (maximum spread). This is in accordance with three different concepts used in these evaluations, namely *determinisity* (deterministic estimate), *probability* (probabilistic estimate), and *possibility* (fuzzy and interval estimate).

The degree of uncertainty tends to vary during the different phases of development. For example, the uncertainty interval for the RF narrows as the project moves from pre-drilling phase toward the actual recovery after drilling. Appropriate methods used in this regard are analogy to other known discoveries, volumetric estimates, production performance analysis, simulations, material balance calculations, and decline curve analysis.

Table 1.2: Results of reserve estimate by using different techniques (in million Sm^3).

	Minimum	Maximum	Most likely	P_{90}	P_{50}	P_{10}
Deterministic	78	11,435	1,121	2,803	2,803	2,803
Probabilistic	78	11,435	1,121	928	2,695	6,865
Fuzzy	78	11,435	1,121	912	3,170	7,525
Interval	78	11,435	5,757[a]	1,214	5,757	10,300

[a]Note that the most likely value in the interval evaluation is simply a middle point between the minimum and maximum reserve values, which substantially differs from other estimates performed by using other techniques.

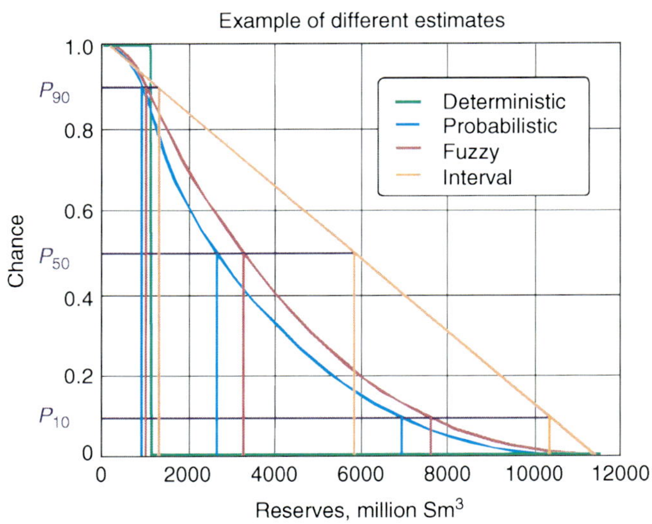

Figure 1.9: Results of the reserve estimate for the given example performed by using different approaches.

1.5 Data acquisition in exploration

To find and produce hydrocarbons requires a detailed knowledge of the earth's subsurface, the rock types, their formation, evolution, geometric shapes, and both past and present relationships.

Many disciplines are brought together to optimize our knowledge, and many different tools are employed: geology, geophysics, physics and chemistry, navigation and mapping, statistics, computer science and economics, engineering and mathematics and, importantly, imagination, optimism, and persuasion.

An unexplored basin has no wells, neither seismic nor gravity data. For an onshore area, it is possible to find outcrop rock samples, which is highly unlikely for offshore areas. As our knowledge of the basin increases through drilling, geophysics continues to mould our geological concepts, and plays a vital role in locating appraisal and development wells. The tracks of deviated wells and horizontal wells can be accurately planned in order to optimally test parts of the prospect or reservoir. The same tools are increasingly being used at mature fields, playing important roles in collecting the data for long-term production planning and economic evaluation.

The process of data collection, analysis, and model evolution is an iterative process. When a field is discovered, more data are gathered using remote-sensing techniques and subsurface sampling. The model is refined, and our understanding improves although the model will always be approximate. Even a mature field may be poorly understood. New information helps to upgrade the model and our understanding of the field. Sometimes the data collected might change the initial model to such an extent that it can result in a termination of the operations and even the abandonment of the field.

1.5.1 Geophysical data

For any part of the earth's subsurface today, information is available to allow some preconception of the geology. Global tectonics will almost always provide some hints as to what might be expected. Onshore, satellite images can provide a surprising insight into the subsurface geology. Surface geology and shallow boreholes, if not already available, can be done quickly and cheaply. However, such data are entirely inadequate to identify or quantify prospects 2 km beneath the surface.

The fastest and cheapest addition to a database of an area is often an *aero-magnetic survey*. Such surveys are based on recording variations in the earth's magnetic field which reveal certain information on structures in the underground. The survey can be recorded from an airplane and processed within weeks. Although poor in resolution, it can indicate the depth to basement, or sedimentary cover, which is a very important parameter for exploration. It can also reveal the structural grain or lineament and the positions of major fault complexes. The aero-magnetic survey can be supplemented with a *gravity survey*, which can also be recorded from an airplane. The integrated results can indicate gross density contrasts in the sedimentary overburden and the presence of salt-induced diapiric structures.

Before even contemplating drilling a well, a grid of seismic data must be acquired and interpreted. Seismic surveying is based on sending shock waves to the underground, recording and interpreting the reflections. Seismic surveying might commence with a cross of just two regional lines that extend across the basin from an outcrop position on the basin's edge. After the interpretation results of this minimal grid, a regional 2D survey might be designed with spacing between lines of, perhaps, 5–10 km in the dip direction and double that in the strike direction. At this stage, some concept of the structural shape and depths of possible prospective formations will be emerging, and subsequent seismic lines can be concentrated in what are considered to be prospective areas. There are conflicting concerns of cost vs. an adequate exploration program. Areas may too easily be written off as being too deep or over/under-mature, etc.

As knowledge of a basin improves, attention becomes focused. Infill (2D) seismic can be recorded to confirm a structural closure or seismic anomaly. 3D seismic is expensive, and it is unlikely that it will be considered economic before there is some subsurface confirmation of the stratigraphy and lithology.

1.5.2 Seismic survey design and acquisition

The cost of seismic acquisition on land is controlled by the accessibility, terrain, and vegetation as well as the cultural environment. On land, using dynamite to create the shock waves provides the best wavelet shape, but its use is precluded in built-up areas. Using vibrations to create the shock waves (*Vibroseis*) is less intrusive, but provides lower data quality.

The line spacing needs to be determined by the complexity of the subsurface structure and the minimum size of anticipated prospect. A 10 km × 10 km grid, for example, would have a poor chance of delineating a closure with an area of 20 km^2.

The design of a 3D seismic grid is subject to all the factors that influence the design of 2D seismic, but with the addition that the line spacing is much denser. Optimal line spacing is influenced by the maximum geological dip and maximum desired signal frequency, similar to how the group interval is constrained in a 2D survey.

1.5.3 Seismic data processing

The objective of data processing is to enhance the wanted signal and attenuate the random and coherent noise. Noise is an inevitable consequence of the seismic recording process. Coherent noise is any stray energy that is organized in some way. It may arise from passing road traffic or rig activity, or may occur as converted surface waves that are always part of the acoustic signal in a layered medium.

The processing includes wavelet shaping, spectral filtering and deconvolution, analysis of the seismic velocities, stacking and migration of the energy. The ultimate objective is a cross-sectional display that portrays the layered sequences.

It is important to realize that the seismic section is not a geological cross-section. A geological cross-section is a snapshot of a vertical cut through the earth; it shows a range of depths at a single instant of time. By contrast, a seismic section shows a range of time at a single depth (the surface). It is not a cross-section, since any boundary reflection is seen through all the overlying layers, which distort the true image.

1.5.4 Subsurface data

Useful interpretation of geophysical data can be obtained without recourse to well data, and this is normally required in the beginning stage of exploring a new basin. The reflection amplitudes from seismic surveys and their geometric relationships can help to quantify structural leads, identify stratigraphic leads, and provide clues on general stratigraphy and gross lithology.

However, a well will be required in order to prove the nature of the formations. Downhole surveys such as check shots and vertical seismic profiling (VSP) will give geological meaning to the reflections seen on the seismic profile. It is important to realize that formations are not homogenous and isotropic, and 100 m away from the well lithology can be significantly different. A well, providing 1 m^2 of detailed information, complements the 100 km^2 of fuzziness provided by seismic.

Cuttings recovered while drilling are analyzed for hydrocarbon shows, oil staining, and paleontological content. A well program may include *measuring while drilling* (MWD) and a suite of wireline logs, perhaps RFT measurements, sidewall cores and hole coring, and perhaps a testing program before being plugged and abandoned or shut in. All these tools are designed to measure the physical parameters of the rocks so that their hydrocarbon-bearing potential may be assessed.

The Check Shot/VSP program is not designed to analyze the rock but to provide a link between well and seismic. A series of travel times from the surface to a known depth in the well are measured and form the basis of a depth–time relationship. This is combined with the sonic log to make a synthetic seismogram, which is compared directly with the seismic profile.

1.5.5 Geological and geophysical interpretation

The objective of geophysical interpretation is to identify some attribute of the available data that might imply an economic accumulation of hydrocarbons. There are historical instances where this has been achieved on the basis of well logs alone, or surface topography alone, but it becomes increasingly necessary to employ sophisticated technology to discover the ever more subtle traps, requiring the use of all possible data. A prospect recommendation will usually be an integrated interpretation of all available data. A brief review of how well data can contribute is given below.

The various downhole tools are used to measure an assortment of basic rock properties:

Logs: Porosity, permeability, density, velocity, resistivity, neutron and gamma activity, bed dip, mineral content, grain size, and sorting.
Cuttings: Hydrocarbon shows, fossils, and minerals.
Repeated formation tests (RFTs): Fluid content, pressures, permeability boundaries, and total organic content.
Cores: Palaeontology, palynology, facies, lithology, mineralogy, fluids, and all physical properties.

These data are interpreted for information on sedimentology and facies, the depositional environment, erosional hiatus, diagenetic history and fluid movement, and evidence of hydrocarbons. A grid of several wells allows a more complete interpretation than a single well. Maps may be constructed at each important stratigraphic level illustrating the distribution of facies and the palaeogeography. It should be remembered, however, that wells are normally drilled on anomalous geological locations – on a structural crest, or adjacent to a fault, where the conditions may not be typical, or even representative of the basin norm.

The interpretation of seismic data can extend our knowledge laterally and below the total depth of the well. The clarity depends on the lateral and temporal frequency of the data, which is reduced with depth. A minimum vertical resolution of about 20 m is the best that can be expected at depths of about 1000 m, and this will decrease to 50 m at 3000 m depth. Within these constraints, a three-dimensional picture may be constructed from the combination of seismic and well data that may include:

- structural shape of identified stratigraphic boundaries;
- depth and faulting in three dimensions and their evolution through time;
- major unconformities and sequence boundaries;
- depositional features such as deltas, slope and basin floor fans, incised channels, and condensed sequences.

Lateral lithologic progression and in some cases fluid content may be deduced using seismic attributes and amplitude vs. offset (AVO) analysis. These features are mapped at specific stratigraphic or formation boundaries. Isopach maps can be derived to show basin development. Fault plane profiles will show formation relationships across the fault.

The arrangement of faults in profile and map views that are interpreted from the seismic data can lead to an understanding of the tectonic style, which helps predicting play types and prospect locations. The timing of the fault movement and tectonic style can often be compared with other areas, and examples from known hydrocarbon producing areas can sometimes be used as analogs.

1.6 Summary

The exploration for oil and gas involves great uncertainty. On the other hand, the need for oil and gas keeps increasing, which creates a need for more exploration and methods to "see into the ground." In this section, we have discussed the basics of petroleum exploration and have considered several methods of estimating resources and recoverable reserve. Results of such estimates serve as a basis for deciding on further investment in a field.

References

Magoon, L.B. & Dow, W.G. 1994. The petroleum system. In: Magoon, L.B. & Dow, W.G. (Eds.), *The Petroleum System – From Source to Trap: AAPG Memoir*, Vol. 60, pp. 3–24.

Zolotukhin, A.B., Gudmestad, O.T., Ermakov A.I., Jakobsen, R.A., Michtchenko, I.T., Vovk, V.S., Løset, S. & Shkhinek, K.N. 2000. *Basics of the Offshore Oil and Gas Field Development and Construction of Marine Facilities in the Arctic*. Oil and Gas Publishing House, Moscow.

Reservoir and production engineering

At the initial stage of field development planning, there is a need to gather as much information as possible to delineate the reservoir and to provide data for its characterization. Sources of information are the regional geology, seismic surveys, and core samples retrieved from wells drilled into the reservoir and its vicinity, and well logs. Interpretation, integration, and reconciliation of the various data must answer the question: is the field likely to produce enough oil and/or gas to provide a reasonable rate of return over the lifetime of the venture?

This chapter addresses how oil and gas behave in reservoirs, and how this behavior can be influenced in order to optimize production. Topics include reservoir rocks and fluids, reservoir dynamics, reservoir characterization and modeling, reservoir recovery and improved oil recovery (IOR) methods, production forecasting methods, and reservoir management during field development and production.

2.1 Reservoir rocks and fluids

Understanding how fluids flow through reservoir rock is essential for reservoir management and recovery calculations. In the previous section, we defined a reservoir as a subsurface body of rock having sufficient porosity and permeability to store and transmit fluids. Porosity and permeability are the key parameters characterizing a reservoir rock. Permeability is again a subject of rock properties and fluid properties, which mostly depend on pressure and temperature. The impregnated volume is a factor here, as well as the saturations. Petrophysics is a study of the physical properties of rocks.

2.1.1 Rock properties

The main reservoir rocks consist of sandstones and/or carbonates. These are sedimentary rocks, which originally formed at the earth's surface from debris (mineral, animal, and vegetable) or by chemical precipitations. They are stratified in successive beds (Figure 2.1).

2.1.1.1 Sandstone reservoirs

Sandstone reservoirs account for 80% of all reservoirs and 60% of oil reserves. Sandstones are composed mainly of quartz grains (silica, SiO_2). In most cases, they exist in a mixture of other rocks like shale and carbonate. Depending on percentage composition, the mixture is called shaly sandstone, carbonate sandstone, etc.

Sandstones are often stratified in a simply superimposed pattern, or with intersecting beds. This results from successive depositions at the shore-line or in the form of fluvial or deltaic alluvia. A vertical cross-section generally shows alternating deposits of sands, shaly sands, silts, and clays or *shales*.

Figure 2.1: Sediments of Jurassic and Triassic origin, Utah, USA. Photo: W.R. Hamilton, United States Geological Survey.

2.1.1.2 Carbonate reservoirs

Carbonate rocks (Figure 2.2) consist of limestone ($CaCO_3$) and/or dolomite ($CaCO_3$ and $MgCO_3$), and often display reservoir characteristics. The high level of heterogeneity in carbonate reservoirs has made managing such assets very challenging in all aspects: from determining the porosity types, connectivity, and fluid flow behavior to well placement and well stimulation.

Carbonate rocks are of varied origins:

- *Detrital*: formed of debris (grains of limestone, shells, etc.).
- *Constructed*: of the reef type.
- *Chemical*: formed by the precipitation of bicarbonate, originating in marine mud.

Shaly carbonates also exist, but "marls," which contain between 35% and 65% shales, do not exhibit reservoir characteristics. This happens because a small proportion of shale, binding the grains together, considerably decreases the permeability.

2.1.1.3 Fluid characterization

The compositional characterization of naturally occurring hydrocarbon mixtures is an integral part of routine petroleum laboratory analysis. Measurements of the gas and liquid phase compositions are done by separate gas chromatographic set-ups with analysis times of up to 1 hour to provide classification of the eluted peaks into alkanes, *iso*-alkanes, naphthenes, and aromatics as well as to determine their concentration.

Figure 2.2: Devonian limestone, Salmon Fork Black River, Alaska, USA. Photo: E.E. Brabb, United States Geological Survey.

The terms alkanes, *iso*-alkanes, naphthenes, and aromatics refer to the structure of carbon atoms in the hydrocarbon molecules. *Alkanes* (*paraffins*) are molecules composed of carbon and hydrogen atoms, where the carbon atoms are arranged in chains without branches or circular closures. Their molecular composition conforms to the following general formula: C_nH_{2n+2}, for instance, propane (C_3H_8). *Iso-alkanes* (*iso-paraffins*) conform to the same general formula, but have branched structures in the arrangement of carbon atoms. *Naphthenes* (*cycloalkanes*) have circular arrangements consisting of five or six carbon atoms, conforming to the general formula C_nH_{2n}. *Aromatics* also have circular arrangements of six carbon atoms, but in a more dense (double-bonded) fashion, for instance, benzene (C_6H_6). Large (heavy) hydrocarbon molecules often cannot be classified clearly, because they tend to feature different parts that may have alkane, aromatic, etc., structures with the same molecule.

The extended compositional analysis throws light on the distribution of the different types of molecules in the reservoir fluid (see Figure 2.3) and provides accurate physical properties for each cut, thus improving the understanding of the thermodynamic and rheological behavior of the petroleum fluids. Hydrocarbons can be classified as dry gas, wet gas, gas condensate, near-critical fluids, volatile oil, black oil, and heavy oil.

2.1.2 Capillary pressure and wettability

Knowledge of the water–oil capillary pressure saturation relationship is necessary for many reservoir engineering tasks. Wettability is the preference of a solid to contact one liquid or gas (the wetting phase), rather than the other. Capillary pressure is a pressure difference between a non-wetting phase and a wetting phase. Since in the majority of cases the water is a wetting

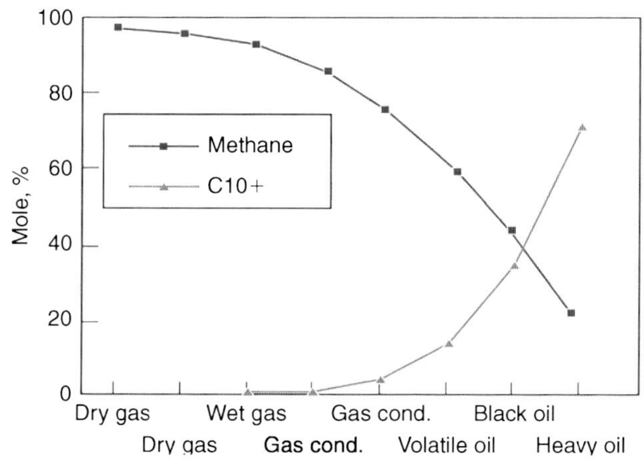

Figure 2.3: Fluid characterization using mole percentage.

phase and the oil is a non-wetting phase, it is suggested to define the capillary pressure as a difference between the pressure in the oil phase and the pressure in the water phase, for example:

$$P_c = p_o - p_w$$

Capillary pressure P_C increases when the wetting fluid is drained out and decreases when the wetting fluid is imbibed (absorbed). When the pressure difference is zero, no fluid movement is occurring (see Figure 2.4). By injecting a wetting fluid, we enhance its pressure and hence enhance recovery due to further imbibitions.

The wetting phase tends to spread on the solid surface, and a porous solid tends to imbibe (absorb) the wetting phase, in both cases displacing the non-wetting phase. Both water and oil wet most materials in preference to gas, but gas can wet sulfur, graphite, and coal.

Drainage (see Figure 2.4(1)) is a process where a wetting fluid (water) is displaced by a non-wetting fluid (oil). It is the same process occurring at the time of oil/gas accumulation in a porous rock that was initially saturated with water.

Imbibition (see Figure 2.4(2)) takes place when water comes in and displaces oil/gas from a reservoir. *Spontaneous imbibition* is a when a process occurs naturally and a *forced imbibition* is when a wetting fluid is injected to enhance recovery.

The primary drainage curve (Figure 2.4) is used to calculate oil in place, and the secondary imbibition curve is used to model oil displacement by both aquifer influx and water injection. By a core analysis, full drainage and imbibition curves for capillary pressure can both be measured, and the capillary pressure curves as a function of fluid saturations can easily be determined.

The wettability affects the relative permeability, electric properties, nuclear magnetic resonance relaxation times, and saturation profiles in the reservoir. The wetting state impacts waterflooding and aquifer encroachment into a reservoir. Reservoir wettability preference can be determined by measuring the contact angle of crude oil and formation water on silica or calcite crystals or by measuring the characteristics of core plugs in either an Amott imbibitions test or a *United States Bureau of Mines* (USBM) test.

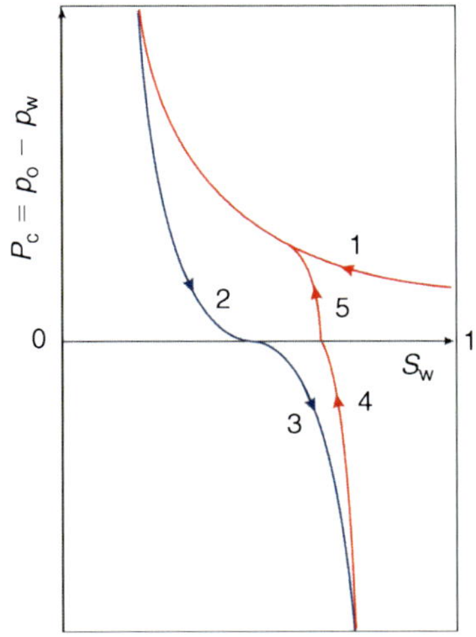

Figure 2.4: Capillary pressure as a function of water saturation. (1) Primary drainage where the wetting fluid (water) is forced out of the rock by the non-wetting fluid (oil); (2) spontaneous imbibition curve of the wetting fluid (water) as the wetting fluid naturally enters the rock; (3) forced imbibition of the wetting fluid (water) as the wetting fluid is injected under pressure into the rock; (4) spontaneous secondary drainage of the non-wetting fluid (oil) as the non-wetting fluid enters the rock; (5) forced secondary drainage as the non-wetting phase (oil) is injected under pressure into the rock.

Permeability is called an *absolute permeability* if a singe fluid or phase is present in the rock. Usually absolute permeability is referred to a porous media when an inert fluid flows through it. Determination of the absolute permeability is usually performed during the laboratory tests on core samples with inert gas flowing through the core. Measured by this method, gas reservoir permeability is then used to calculate the true (liquid) absolute permeability of a core sample. Such correction is needed in order to account for a so-called Klinkenberg effect (Klinkenberg, 1941). At relatively low pressure, gas tends to behave like an inviscous fluid, which means negligible friction against the pore-channel walls. The gas molecules are very small, and the low pressure reduces their mutual collisions and allows them to move through a pore channel with minimum frictional interaction with the walls. As a result, the flow velocity at the wall is non-zero and the bulk flow velocity is effectively higher than might be expected for a liquid or a high-pressure gas.

Phase (or effective) permeability is the ability of a particular fluid (phase) to preferentially flow through a rock when other immiscible fluids are present in the reservoir (e.g., we use the term phase (effective) permeability of gas in a gas–water reservoir). The fluid saturations as well as the nature of the reservoir affect the phase permeability. *Relative phase permeability* is the ratio of phase permeability of a particular fluid at a particular saturation to absolute permeability of that fluid at total saturation. If a single fluid is present in a rock, its relative permeability is 1.0. Calculation of relative permeability allows for comparison of the abilities of different fluids to flow in the presence of each other, since the presence of more than one fluid generally inhibits flow. Relative phase permeabilities are functions of the respective saturations, as illustrated in Figure 2.5.

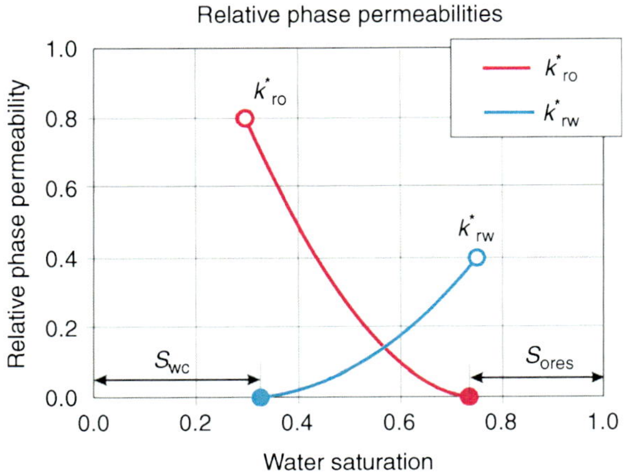

Figure 2.5: Typical relative phase permeability curves for the oil–water system in a water-wetting reservoir. S_{wc}: connate water saturation; S_{ores}: residual oil saturation; k_{ro}^*: end-point relative permeability of oil; k_{rw}^*: end-point relative permeability of water.

Note that relative phase permeabilities are functions of the respective saturations and depend on the wettability preference, direction of flow (drainage and imbibition), and presence of chemicals or other additives (microbes) in different phases.

2.1.3 Hydrocarbon PVT and phase behavior

Pressure–volume–temperature (PVT) analysis of fluid inclusions that are trapped in source rocks, migration pathways, and reservoirs is a powerful tool for quantitatively estimating paleopressures, paleotemperatures, and changes in fluid compositions over time.

Hydrocarbon fluids comprise a multitude of constituents ranging from single-carbon methane to very long-chain carbon compounds, as well as cyclic, aromatic, and other complex molecules such as asphaltenes and waxes. These constituents determine the phase behavior of a given reservoir fluid, which is often indicated using a PVT phase diagram. Near the bubble point curve, the gas phase is predominantly methane, but further into the two-phase region, more light and intermediate components enter the gas phase. Similarly, the first liquid components to drop out after passing the dew point are the heavier components; lighter components go into the liquid phase at conditions further from the dew point curve.

Reservoir temperature is usually almost constant, unless cold or hot fluids are injected into the reservoir, so most depleting reservoirs follow a downward vertical path on a pressure–temperature phase diagram. If the reservoir temperature is between the critical point and the cricondentherm (C_T), liquid can drop out of the gas phase within the reservoir. These reservoirs are called gas-condensate, or retrograde-condensate, reservoirs. Gas in a reservoir with a temperature greater than the cricondentherm is termed a *wet gas* if liquid drops out because of pressure and temperature decreases in the production system, or it is called a *dry gas* if no liquid falls out in either the reservoir or the production system.

2.1.3.1 Formation volume factors and solution gas–oil ratio

Formation volume factors (B_o, B_g, and B_w) for oil, gas, or water are defined as a ratio of volume at reservoir conditions divided by the volume at standard conditions. Since most measurements of oil and gas production are made at the surface, and since fluid flow takes place in the formation, volume factors are needed to convert measured surface volumes to reservoir conditions. Oil formation volume factors are almost always greater than 1.0 because the oil in the formation usually contains dissolved gas that comes out of solution in the wellbore with dropping pressure (see Figure 2.6).

In addition to formation volume factors described above, there are two more important parameters that influence production performance:

R_S *is the gas/oil ratio (GOR).* It is the ratio of produced gas to produced oil. It remains constant at pressures above the bubble point pressure if water is injected. It rises sharply at pressures above the bubble point pressure when a breakthrough of injected gas in the production well occurs. When the bottomhole pressure falls below the bubble point pressure, GOR can sharply increase and stand high until the gas liberated from the oil phase is produced. When the oil is almost free of dissolved gas, GOR sharply decreases and the oil production ceases, as shown in Figure 2.7.

R_V *is the condensate/gas ratio (CGR).* It is also referred to as condensate yield. It is the ratio of produced condensate to produced gas (CGR). It remains constant at pressures above the dew point pressure.

2.2 Reservoir characterization and in-place volumes

To understand and correctly describe the reservoir architecture is one of the most challenging and important stages of field development. It serves as a basis for evaluating initial and recoverable volumes of hydrocarbons, required number of wells, their type, completion design, and the drainage strategy.

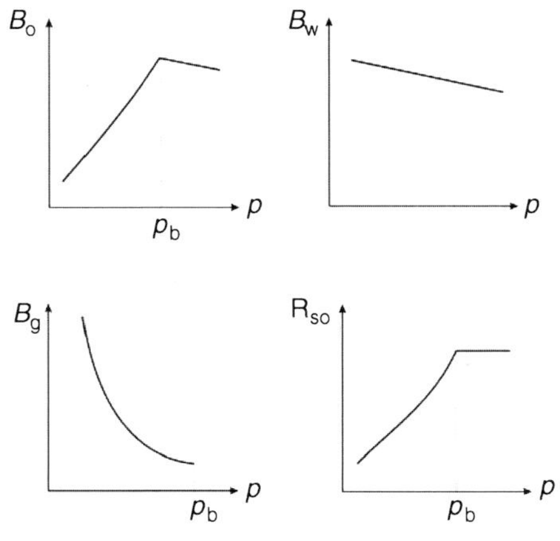

Figure 2.6: Effect of pressure on formation volume factor for oil, gas and water.

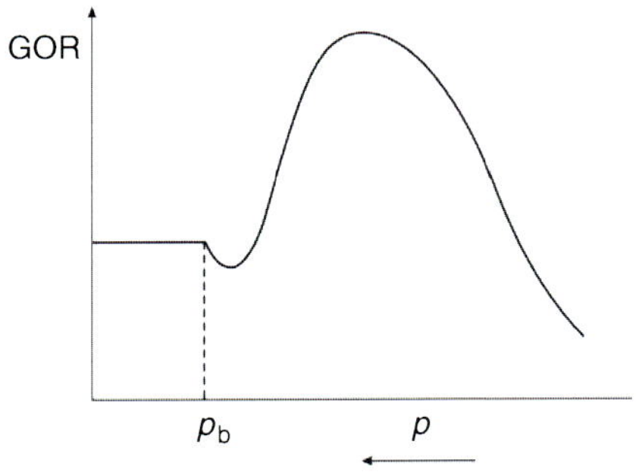

Figure 2.7: Typical behavior of produced GOR during pressure depletion.

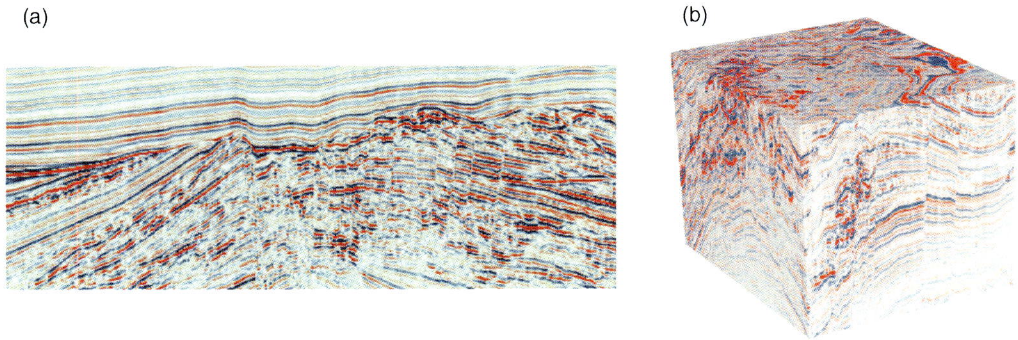

Figure 2.8: Seismic profile (a) and seismic cube (b) as a result of 2D and 3D seismic data interpretation.

The reservoir model is constructed from incomplete data, and much of the inter-well space must be estimated from information from nearby wells or from low vertical resolution data, such as seismic data. The construction of models can be performed by deterministic methods or geostatistical methods, or by a combination of both.

Recent advances in geology, geophysics, and computer science enable construction of very sophisticated and much more realistic models than those used in the 1990s. Figure 2.8 shows examples of seismic interpretation of 2D (a) and 3D (b) seismic data. A so-called *seismic cube* (b) provides more detailed data and enables more accurate estimate of reservoir volumes as compared with *seismic profile* (a).

This information coupled with well data gives the basis for the detailed reservoir characterization that, in turn, enables more accurate resources and reserve estimate and further reduces uncertainty in forecasting reservoir performance. Moreover, reservoir heterogeneity can be modeled in three-dimensional space, which helps to evaluate well and reservoir production characteristics.

Estimation of in-place volume (OIIP or STOIIP) can be carried out directly by volumetric analysis based on the available reservoir model(s), or indirectly by *material balance* methods.

Volumetric analysis and evaluation of accompanied uncertainties has in detail been covered in a previous (geology) section where three different approaches were discussed.

Material balance methods provide a *dynamic* measure of hydrocarbon volumes. The accuracy depends on the quality of data for: *pressure surveys*, *temperature surveys,* and *analysis of recovered fluids*. Normally accuracy of our estimates by mass balance methods increases when more data on production performance become available.

2.3 Factors affecting reservoir performance

Among the factors affecting reservoir performance the most important are listed as follows:

- reservoir geometry and heterogeneity;
- unfavorable (adverse) mobility ratio between the *displacing* (water and gas) and *displaced* (oil) fluids;
- number of wells, and their type and placement (drainage strategy);
- recovery method and recovery mechanism;
- water and gas coning effects.

The first three factors lead to low volumetric (macroscopic) sweep efficiency, one of the two components of the oil recovery factor. Other factors leading to an insufficient recovery are the displacement efficiency of water (often called microscopic sweep efficiency) and the difference in densities of the displacing (water and gas) and displaced (oil) fluid.

2.3.1 Oil entrapment

Oil entrapment often occurs when waterflooding is performed in heterogeneous formations. This effect can be even more considerable when the reservoir contains viscous oil, resulting in an unfavorable mobility ratio between the displacing and displaced fluids. This effect is shown as a simulation result in Figure 2.9.

2.3.2 Low sweep efficiency

As has been discussed above, a combination of the reservoir heterogeneity, unfavorable mobility ratio, and difference in densities of the displacing and displaced fluids leads to low sweep

Figure 2.9: Oil entrapment (Zolotukhin & Ursin, 2000).

efficiency and a poor production performance. Figure 2.10 shows the results of a stochastic simulation of waterflooding in a core sample (a) and in a well pattern containing one injector and one producer (b). Both simulations indicate low sweep efficiency of waterflooding at the breakthrough time due to adverse mobility ratios.

2.3.3 Water and gas coning

Differences in densities of gas, oil, and water often lead to a coning effect when the gas (injected or from a gas cap) moves downward and the water encroaches upward toward the producing well (Figure 2.11). Due to the higher mobility of both gas and water as compared

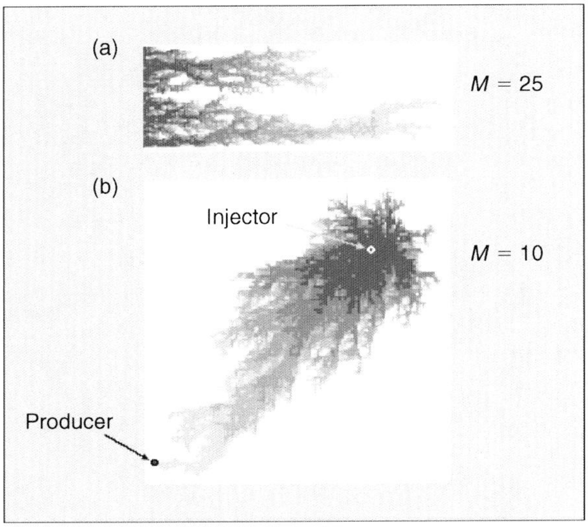

Figure 2.10: Stochastic simulation of waterflooding in a core sample (a) and in a well pattern containing one injector and one producer (Zolotukhin & Ursin, 2000). Dark shades indicate large water saturation.

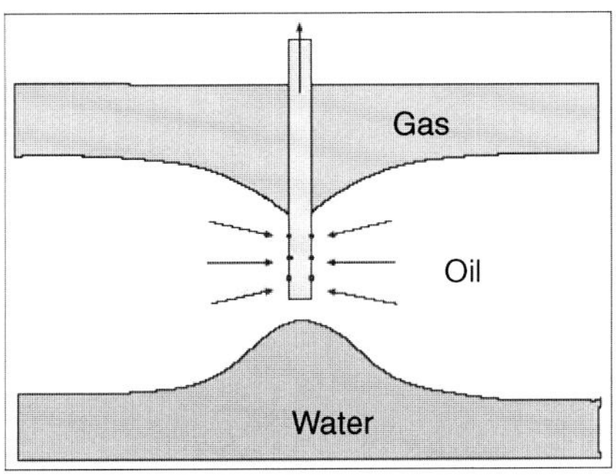

Figure 2.11: Water and gas cone formation.

to that of oil, this effect results in a low volumetric sweep efficiency and poor production performance of producing wells (high GOR, high water cut, and low oil production rate).

2.3.4 Number of wells, their type, and placement (drainage strategy)

The total number of wells, their type (injector and producer), design (vertical, slant, horizontal, fractured, etc.), and placement determine a *drainage strategy*, which greatly affects production performance.

Injection wells are used for maintaining reservoir pressure at a sufficiently high level to secure high production rates, and for improving and maintaining a good macroscopic sweep. High reservoir pressure prevents a field from developing a solution gas drive. Injection drive gives a good production performance while solution gas drive results in a poor production performance.

Re-injection of produced water and gas (associated gas or sequestrated CO_2) into aquifers or in hydrocarbon-bearing formations is another reason for drilling injection wells, so as to avoid their release to the surface environment.

Correct evaluation of the total number of wells required for the best reservoir performance is one of the most challenging optimization tasks.

2.4 Methods of hydrocarbon recovery

This section is mostly related to oil recovery methods. While in gas-bearing formations a gas recovery factor often exceeds 70% and is not a big issue, for oil fields the oil recovery rarely reaches 50%. This fact makes the oil recovery a primary concern for detailed reservoir and production engineering.

Studies show that the average recovery efficiency could be closely tied to depositional environment and drive mechanism. Reservoirs composed of fluvial rocks with water drive or gas-cap expansion may have recovery efficiencies of 35–50%, whereas barrier strand-plain reservoirs with gas-cap expansion may have recovery efficiencies of 50–70% (Larue & Yue, 2003).

Depending on the source of energy and time of deployment, different recovery methods can be grouped into *primary*, *secondary*, and *tertiary* recovery methods.

2.4.1 Recovery by pressure depletion

During pressure depletion, there are several physical mechanisms that are active. These are often called *natural drive mechanisms*. They define production of oil from a reservoir by a dominant natural energy (gas cap, solution gas, water drive, gravity forces, compressibility and compaction, etc.) that results in flowing wells, or wells on the pump with the oil flowing freely by gravity to the well bore. The recovery factor from natural drive mechanisms is typically within the range of 15–20%.

The following *natural* drive mechanisms can be revealed under the pressure depletion phase:

- fluid expansion;
- natural water drive;
- solution gas drive;
- gas-cap drive;

- gravity drainage;
- compaction drive.

Fluid expansion during depletion drives the oil from the reservoir toward the production well due to the compressibility of oil and water in the pores. Oil recovery due to fluid expansion developed within an oil zone is usually less than 2%.

If an oil reservoir has an aquifer (extended water zone) with good permeability, then a *natural water drive* is exhibited. Natural water drive can give a recovery factor of 35–75%.

When the reservoir pressure drops below the bubble point, pressure solution gas previously dissolved in water appears as a free phase. When pressure drops further, there is an expansion of highly compressible gas bubbles, expelling the oil from porous media (*solution gas drive mechanism*). Solution gas drive lasts longer in stratified reservoirs where gravity segregation of gas is deteriorated due to low average vertical permeability. In some cases, gas previously dissolved in oil moves up-dip, forming a so-called secondary gas cap and replacing attic oil. On average, solution gas drive is the least efficient mechanism of oil recovery, as only 5–25% of the oil initially in place can be produced.

When a reservoir has a gas cap, a *gas-cap drive* is always exhibited. High gas compressibility and the extended gas cap size ensure a long-lasting and efficient field performance due to a gas-cap drive: up to 40% of the original oil in place can be recovered under a gas-cap drive.

In formations with a high dip angle, *gravity forces* may contribute significantly to oil recovery. In some cases, *gravity drainage* might result in extremely high oil recovery factor. However, this tends to be an uneconomic drainage strategy due to low well rates.

Compaction drive might occur during depletion when rock grains are subjected to stresses (i.e., the difference between the overburden and pore pressure) beyond the elasticity limit. It leads to a recompaction of partially deformed or even destroyed rock grains that, in turn, might result in gradual or abrupt reduction of the reservoir pore volume. While the former often is accompanied by a subsidence phenomenon, the latter may lead to local but destructive earthquakes.

2.4.2 Recovery by injection

Recovery by injection gives primarily pressure support and thereupon sweeps support, depending on the geometry, layering, and aquifer size. The recovery factor becomes typically 15–45%.

If the natural drive mechanisms in an oil reservoir do not provide adequate energy for oil recovery, secondary recovery operations can supplement natural energy by injection of water and/or gas.

In terms of recovery and cost, the single most important decision to be made in a development is whether to inject water or gas. Injection should improve recovery, but the associated costs with injection equipment and facilities for handling and disposing the injected fluid when it enters the production line need to be economically justified. To make a decision regarding this issue is mainly a risk in case of an offshore development since the numbers and sizing of the installations are strongly influenced by the injection options selected. In an onshore development, it is usually prudent to gather production information over time and establish a positive cash flow, before embarking on an injection scheme.

2.4.2.1 Water injection

Water injection is the most conventional choice of injection fluid, which is due to the abundance and cost. Its efficiency as a displacing medium is dependent on how large reservoir volumes it contacts (volumetric sweep efficiency) and how much of the contacted oil it is able to mobilize (microscopic sweep efficiency). The mobility ratio is a major determinant of waterflooding efficiency. This fact is illustrated in Figure 2.12 showing the effect of the mobility ratio on the volumetric sweep efficiency.

In some of the existing classifications, the methods described below are commonly named as enhanced oil recovery (EOR) methods.

Main objectives of EOR methods can be specified as follows:

- *Enhanced displacement efficiency*: Some of the agents injected into the reservoir (solvents, surfactants, etc.) reduce residual oil saturation and, thus, enhance the oil displacement efficiency.
- *Improved sweep efficiency*: Technologies like polymer flooding, water-alternating gas injection (WAG), foam flooding, thermal methods of oil recovery, etc., aim at improving the *mobility ratio M* and thus, *sweep efficiency*.

Mobility ratio is defined here as the ratio of mobility of the displacing (water and gas) and the displaced (oil) fluids at end-point relative phase permeabilities (Figure 2.5):

$$M = \frac{k_{\mathrm{rw}}{}^*}{\mu_{\mathrm{w}}} \frac{\mu_{\mathrm{o}}}{k_{\mathrm{ro}}{}^*} = \frac{k_{\mathrm{rw}}{}^*}{k_{\mathrm{ro}}{}^*} \frac{\mu_{\mathrm{o}}}{\mu_{\mathrm{w}}} \tag{2.1}$$

where M is the mobility ratio; $k_{\mathrm{rw}}{}^*$ the end-point relative permeability of water; $k_{\mathrm{ro}}{}^*$ the end-point relative permeability of oil; μ_{o} the viscosity of the oil; and μ_{w} the viscosity of the water.

When special core analysis data are not available, the mobility ratio can be estimated as a ratio of viscosities of the displaced (oil) to the displacing (water and gas) fluid:

$$M \approx \frac{\mu_{\mathrm{o}}}{\mu_{\mathrm{w}}} \tag{2.2}$$

Conventional (primary and secondary) methods of oil recovery usually result in less than 45% of the recoverable resources. The major portion of petroleum remains in place.

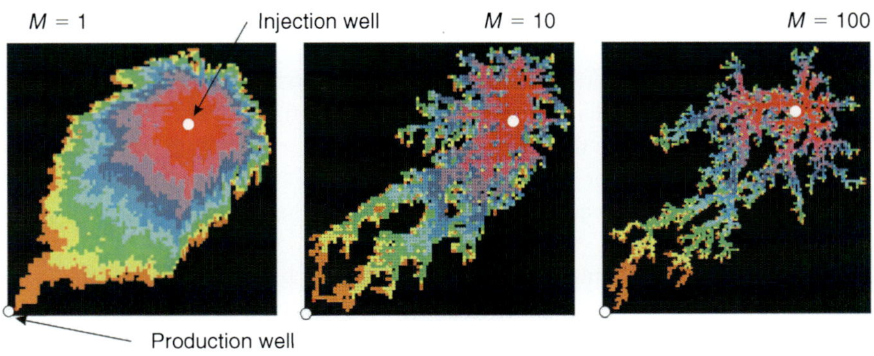

Figure 2.12: Effect of mobility ratio M on waterflooding performance (Frick & Zolotukhin, 1994). Aerial sweep at the water breakthrough time. Different colors represent aerial sweep at different stages of waterflooding; black color represents the unswept oil.

This unrecovered quantity depends on the complexity of reservoir conditions and the field development strategy. The selection of a high-recovery strategy can be constrained by economics, that is, the added expenditures need justification through added revenues at an acceptable rate of return. EOR/IOR methods are effective in upgrading reserves that can be economically recovered beyond conventional recovery methods by augmenting volumetric sweep (macroscopic) efficiency and/or by enhancement of the displacement (microscopic) efficiency.

Various definitions have been extensively used to classify different oil recovery methods. This is in particular true for EOR methods. Applied here is the definition given by the Norwegian Petroleum Directorate (NPD):

> Enhanced Oil Recovery (EOR) refers to advanced recovery techniques going beyond what is considered as conventional methods at a given reference point in time.

Typical features of advanced recovery techniques are injection into the reservoir of other substances than water and/or hydrocarbon gas, and/or that a technical qualification of feasibility is required, for example, a pilot project.

Advanced methods may include:

- WAG;
- chemical methods (surfactant, polymer, gel, foam, etc.);
- injection of gases other than hydrocarbon gas (CO_2, N_2, flue gas, etc.);
- microbial methods;
- thermal recovery methods.

Below is a brief description of some of the EOR methods.

2.4.2.2 Immiscible gas injection

If a hydrocarbon gas is utilized as the injection medium in non-miscible mode, alone or in combination with water, the impact on sweep efficiencies varies with the prevailing situation. Gas, due to its low viscosity, will always have a higher mobility than oil, usually resulting in fairly poor volumetric sweep efficiency due to viscous fingering. On the other hand, due to lower interfacial tension between gas and oil, the residual oil saturation after gas flooding tends to be lower than for waterflooding.

Also gravity plays a bigger part in the context of gas flooding. Oil has a much higher density than the gas. This implies that if gas is injected below the flanks of the oil zone, gravity will move the gas very rapidly to the ceiling of the structure, resulting in very poor volumetric sweep efficiency between injector and producer. However, if attic oil exists above the producer well, a secondary gas cap can displace the attic oil toward the producer well, thereby increasing the recovery.

If gas is injected in the structural high and the reservoir has a reasonable dip, a secondary gas cap can form and gravity can serve to stabilize the descending gas–oil contact, allowing for good volumetric sweep efficiency before the contact is so close to the perforation of the producer that the mobility difference overrides the gravity forces and destabilizes the contact, resulting in gas fingering into the perforation (Figure 2.13).

As shown in Figure 2.13, down-dip gas injection (a) results in forming a secondary gas cap and in favorable displacement scenario, while up-dip gas injection (b) leads to a gas overriding process that deteriorates the reservoir performance.

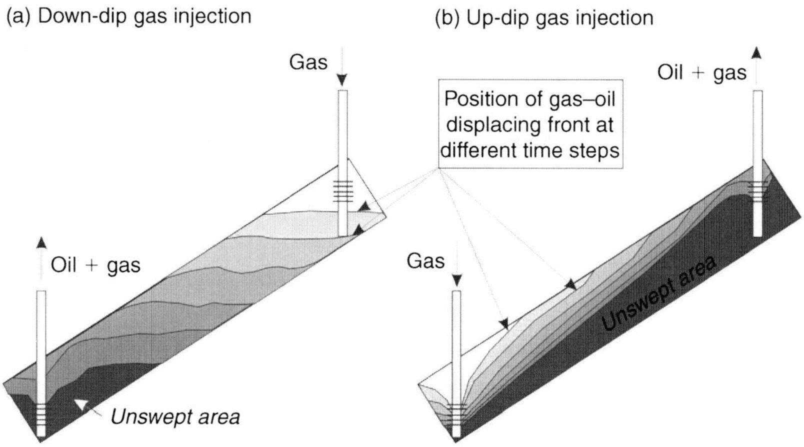

Figure 2.13: (a and b): Effect of gas injection into a dip reservoir (Zolotukhin, 2000).

2.4.2.3 Miscible gas injection

In case of miscible gas injection, the improved recovery mainly stems from low residual oil saturation due to very low interfacial tension between oil and gas and swelling of the oil due to gas being dissolved in the oil. Miscibility is generally considered as a favorable condition, and is commonly evaluated if gas injection is considered.

Justifying hydrocarbon gas injection tends to represent a bigger challenge than justifying water injection, since gas may have an alternative value. Gas injection equipment such as high-pressure gas compressors is expensive. Associated gas used for injection may have a price because it could otherwise be sold, and if there is a need to procure additional gas, this will also have a price. In areas where natural gas is produced but cannot reach a substantial market, the effective price of gas may be very low or zero.

Gas flaring and venting are increasingly regarded as unacceptable practices and may be prohibited by regulations, both for environmental reasons and in order not to waste a potentially valuable resource. Following extensive gas injection, sooner or later free gas will break through in the producer wells, which may trigger a need for costly gas handling and disposal. Gas injection can therefore be a good solution for some time, but when free gas starts to fill up the handling capacities, sales outlet must be found. As a means for gas disposal, injection is only a temporary solution.

2.4.2.4 Other injection fluids

At the end of primary and secondary recovery operations, on a worldwide average, only about 30% of the oil in place is recovered. EOR techniques work on the basis of injecting fluids or chemicals into the reservoir, which alter the properties of the reservoir fluids or the fluid–rock interaction to facilitate the improvement of the recovery factor. Methods deployed are intended to improve volumetric sweep or microscopic sweep as follows:

Improve volumetric sweep:

- Adding polymers to increase water viscosity and improve mobility ratio;
- adding energy to the reservoir to increase *in situ* fluid temperature and decrease oil viscosity, thereby increasing its mobility (this can be achieved by use of steam, hot water, or combustion);

- injecting permeability-reducing or -blocking agents in injectors and/or producers to selectively change the water or gas production or injection profiles (polymers, silica gels, and foams);
- adding another fluid phase, that is, gas, with different physical and chemical properties (hydrocarbon gas, CO_2, nitrogen, and flue gas).

Improve microscopic sweep:

- Adding chemicals to the water phase to reduce the interfacial tension with oil (these chemicals are typically surfactants, alcohols, or caustics, alone or in combinations);
- adding nutrients to enhance bacterial growth, acting in the reservoir as surfactants (MEOR).

As follows from the text, there are two ways of classifying reservoir management strategies. The first is based on the timing of deployment (primary, secondary, and tertiary methods), while the second refers to the method's specific features (natural drive and EOR/IOR).

In all cases, the oil in place is a target to the IOR methods if it can be produced profitably. This implies that the amount of oil that can be recovered by means of IOR methods is subjected to certain conditions (political, economic, technological, etc.) and does not represent a constant value.

2.4.3 Reservoir engineering during operations and improved oil recovery

Several decades ago, the term *IOR* was introduced. It comprises all methods and techniques leading to the increased oil recovery. IOR is the planning and execution of measures intended to increase or accelerate oil recovery from a hydrocarbon field in a profitable manner compared to current plans and forecasts.

Conventional IOR methods include, but are not limited by:

- injection of water and/or gas;
- infill drilling;
- horizontal wells for draining thin oil zones or remaining oil pockets;
- long-reach wells for draining oil in the outer flanks of the reservoir;
- upgrading of treatment capacity for produced water and/or gas;
- reduced wellhead pressure and artificial lift in the wells;
- change of completion strategy;
- tie-in of satellite fields to existing infrastructure and processing facilities, etc.

According to the NPD:

> Improved Oil Recovery (IOR) is a term embracing all methods resulting in an increased oil recovery factor from a reservoir as compared with the expected value at a certain reference point in time.

IOR may be achieved by using conventional methods including improved reservoir management and cost-reducing measures, or by using advanced methods (EOR). IOR can be grouped into *secondary* and *tertiary* recovery.

2.5 Simulation of reservoir dynamics

It is difficult to overstate the importance of reservoir performance simulation. It enables us to understand processes occurring in productive formation, control them, and predict the reservoir performance. Without simulation, it is impossible to maintain the reservoir management and the optimization of hydrocarbon recovery at the required level.

The terms *modeling* and *model* should be kept apart. Usually the term *modeling* means the process of gaining new knowledge about the process, phenomenon, and its translation into mathematical formulation. Modeling can be performed by different means, the most common among them being experimental and mathematical modeling.

The model, to some extent, is a final result of efforts in modeling. However, even in the best case, it is yet an idealized and simplified representation of reality. To study complex problems (e.g., reservoir performance simulation), we usually consider a set or *hierarchy* of models describing different facets of the same problem (or the same features at different scales of observation), thus yielding new information to our knowledge.

Conceptual models used in reservoir engineering can be divided in the following categories:

- models based on principles of continuum mechanics;
- network models;
- micro-models;
- phenomenological models.

2.5.1 Models based on principles of continuum mechanics

According to the principle of *continuum mechanics*, porous media saturated with fluids can be described as *continuum*, that is, media containing all its components (rock + fluids) in any volume selected at random, even if the volume approaches zero. This approach was introduced to reservoir engineering in the beginning of the 20th century and since that time it has been most successfully applied to numerous reservoir engineering tasks. The overwhelming majority of numerical models (ECLIPSE, MORE, FRANLAB, UTCHEM, and many others) used in reservoir engineering are based on this principle.

Based on this approach, equations of continuity (or mass balance equations) can be written for each phase presented in porous media. Coupled with equations of flow (which are mostly phenomenological or empirical equations), and initial and boundary conditions, these equations represent a multiphase (compositional, if several components are considered) flow model that evaluates saturation and pressure distribution in the reservoir and, therefore, enables to control and predict the reservoir performance.

However, these models have a certain limitation: they do not describe phenomena occurring at *micro-scale*, that is, when the volume considered is comparable to the pore or grain size. However, parameters representing micro-scale phenomena appear in continuum mechanics models as constants (connate water saturation, residual oil saturation, etc.) or as certain empirical relations (relative phase permeability and capillary pressure) that cannot be directly derived from these models. In order to obtain these relationships/constants, smaller scales of observation and appropriate models and/or experimental results should be considered.

Description of other models used in reservoir engineering can be found in Zolotukhin and Ursin (2000).

2.5.2 Hierarchy of models

As follows from the previous discussion, various models can be used to study the same process at different scales, as illustrated in Figure 2.14. When examining the mechanism of oil displacement by different agents from a capillary channel (i.e., considering distances compared to a pore-channel size), micro-models can be efficiently used. At the next level, where thousands of porous channels create a certain pattern, the network models can be fruitfully applied in order to obtain important properties of fluid flow, such as relative phase permeabilities, end-point saturations, etc. However, such models are not efficient tools for modeling fluid flow at macro-level when the distances considered are too large (meters and tens of meters), since the detailed information on porous structure (on which micro-models are based) is lost at this scale.

Macro-models allow examining the process of displacement of one fluid by another, mass transfer phenomena at macro-scale (e.g., mass transfer between porous blocks and fractures), distribution of phases and pressure distribution within the reservoir, etc. In order to describe fluid flow at macro-scale, continuum mechanics models and empirical models based on available statistics are mostly used. Their application provides a smoothed picture of fluid flow. However, it enables controlling effectively the growing dimension of the problem, although it is reached through some coarsening of the results.

At mega-scale, when the reservoir performance is considered, some difficulties in using macro-models may arise. Among them the most important are:

- A problem of growing dimension. In some cases (large fields with complex geological structure, complex reservoir/fluid properties, large number of wells, etc.), even the most powerful computers cannot handle the problem effectively. In order to avoid arising difficulties, different techniques for *upscaling* the data are used. As a marginal result of such averaging of reservoir/fluid parameters, information about the reservoir accepted by the model becomes less informative, and the model may lose its predictive power.
- Reservoir characterization at a field scale is still a challenging problem. Often the lack of knowledge is substituted by different hypotheses whose verification can be problematic. Uncertainty in reservoir characterization increases at this scale and, as a result, models may become increasingly formal and less constructive.
- Optimization of field performance is yet another big problem which cannot be efficiently tackled by macro-models.

In order to handle the above problems efficiently, *mega-models* can be used. These models describe the most general and important characteristics of the field performance, such as dynamics of the oil recovery, water–oil ratio, water cut, duration time of filed development, etc. Mega-models can be composed by using empirical relations and field performance data.

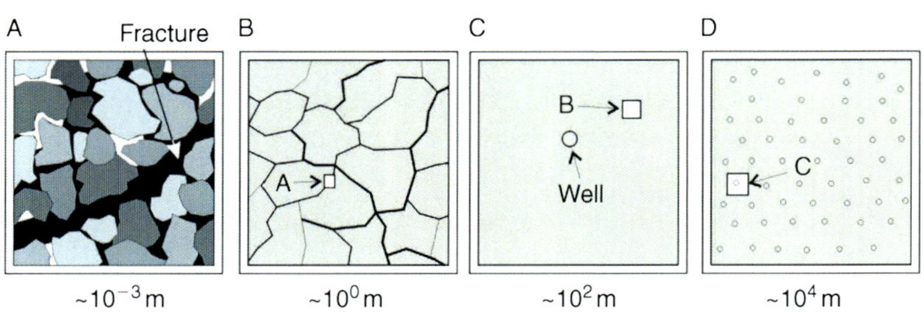

Figure 2.14: Structure of a fractured rock at different scales of observation.

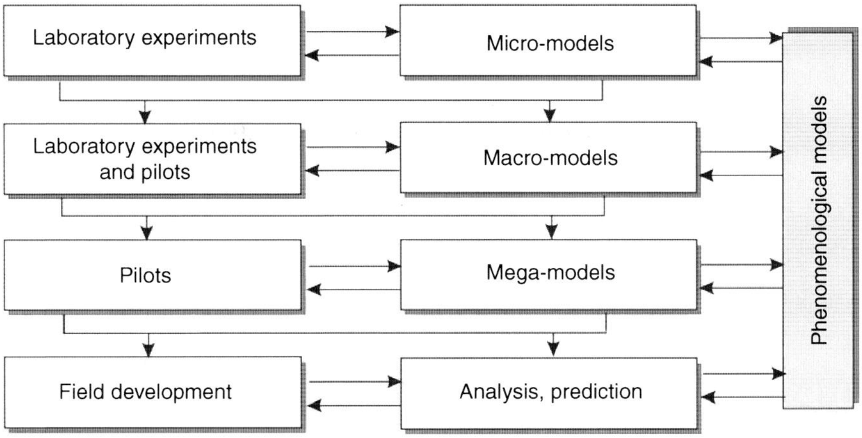

Figure 2.15: Hierarchy of models used in reservoir and production engineering.

Note that models of different levels are related to each other. It means that models of different levels represent a hierarchical structure where every element (a model) is required to describe the process, the phenomenon, at the desired level of detail (Figure 2.15). Moreover, these hierarchical levels are connected to each other: the upper level models consume the information obtained at the lower level. On the other hand, results obtained at the lower level can be used as the input data for the models at the next hierarchical level.

2.6 Production planning and reservoir performance forecast

Understanding the physics of fluid flow through porous media is a necessity for successful production planning and reservoir performance forecast.

So far, numerical simulation is considered the most adequate and accurate modeling of reservoir performance, enabling reliable production forecasting. However, it is not the only tool used for predicting reservoir performance. Due to the complexity of reservoir behavior, a number of engineering methods have evolved depending on assumptions made in analyzing the fluid flow. Among methods available for production forecast the most popular or most often used are the following:

- methods based on reservoir performance simulation (numerical simulation methods and analytical methods);
- methods based on reservoir performance analysis (material balance equation (MBE), decline curve analysis (DCA), and similar techniques used at the depletion stage);
- simplified forecasting (engineering) methods.

Below we briefly describe essentials of some of the most used forecasting methods.

2.6.1 Methods based on reservoir simulation

2.6.1.1 Numerical simulation methods

Numerical simulation is a mathematical simulation using a numerical model of a reservoir's petrophysical characteristics to analyze and predict fluid behavior in the reservoir over time.

It is a sequence of numerical operations whose output represents the behavior of a particular process in a particular reservoir. It is, therefore, an integrating tool for combining, in their proper weight, all the factors that influence production in a reservoir. The model is constructed from incomplete data and much of the inter-well space must be estimated from information from nearby wells or from low vertical resolution data, such as seismic data. The construction of models can be performed by deterministic methods or geostatistical methods, or by a combination of both.

Models are nothing more than solutions to conservation equations that are coupled to several phenomenological laws that are needed to make up the required number of equations and unknowns. Listed below are the basic equations and laws associated with reservoir simulation models based on *continuum mechanics principle:*

- material balance (conservation of mass, energy, and momentum);
- empirical laws: Darcy, capillary pressure, phase behavior, reaction rates, etc.

These equations are far too complicated to be solved analytically (directly). Instead, when solving for a particular reservoir, these equations are divided up into a number of cells or grid blocks and then solved on these much smaller volumes. This division is, of course, arbitrary but it is the only way to proceed. It would be no surprise to find that the division into grid blocks leads to a number of artifacts in the simulator results. But it should equally be of no surprise to realize that these artifacts tend to disappear as the number of blocks increases; with current simulation technology, the artifacts seem to comprise less of a problem than simply not knowing the simulator input initially.

The reservoir simulation modeling involves a numerical simulation technique, such as the finite differences method, the finite elements method, or other techniques used to solve a system of partial differential equations. It predicts a reservoir's performance during production, simultaneously matching it with the reservoir's previous behavior and taking into account all available data.

As a result of simulation, a long-term reservoir production performance can be generated with production profiles for all fluids flowing in porous media. So-called compositional simulation models can even generate production profiles for most important hydrocarbons (dry gas, liquefied petroleum gas, condensate, etc.) or even for their fractions and individual components.

2.6.1.2 Analytical models

Analytical models aim to make a rigorous mathematical and yet analytical description of fluid flow in porous media for special purposes, and to make a simplified but yet analytical prediction of production performance. Analytical models are based on MBEs and set of assumptions enabling the analytical solution. Usually, these models are aimed at a substantial part of a reservoir, the whole reservoir, or an entire field (a mega-scale approach) and, per definition, are based on the *continuum mechanics principle.*

Among analytical models mostly used for predicting reservoir performance are fractional flow models (Buckley–Leverett analysis, Dietz stability analysis, and Dykstra–Parson's method for stratified layers).

An advantage of analytical methods is that they present solution to a given problem in an analytical form, which greatly simplifies further analysis. Analytical models often generate important correlations and dependency curves that can be used in other models, including numerical simulation. In some cases, the analytical approach enables solution of optimization tasks in analytical form – most challenging tasks in reservoir and production engineering. A common disadvantage of these methods, however, is that their area of application is limited by assumptions on flow geometry (one-dimensional flow, linear or radial) and fluid/reservoir properties (homogeneous porous media, black oil model, ignorance of capillary forces, etc.).

Descriptions of the different approaches used in analytical modeling of reservoir performance, and engineering approaches based on them, can be found in numerous books from which we would recommend the following: Amyx et al. (1960), Dake (1978), Archer and Wall (1991), Bedrikovetsky (1993), and Zolotukhin and Ursin (2000).

2.6.2 Methods based on reservoir performance analysis

The aim of methods based on reservoir performance analysis is to assess future reservoir performance on the basis of available data, which may also allow a better estimation of reserves.

These methods describe reservoir performance at macro- and mega-scales (see previous section).

2.6.2.1 Material balance

Basically, a MBE is a statement that accounts for the volumes and quantities of fluids that are initially present in, produced from, injected into, and remain in a reservoir at any state of its depletion. It is an important statement in reservoir characterization to identify drive mechanisms and confirm reservoir volumetric and aquifer fitting. The method was developed into a strong mathematical tool by Havlena and Odeh (1963; 1964). Its numerous applications to reservoir performance forecast based on performance analysis can be found in the original papers of Havlena and Odeh, and in a textbook (Zolotukhin & Ursin, 2000).

2.6.2.2 Decline curve analysis

DCA predicts the reservoir's future performance by matching the observed trend of the production decline with one or several standard mathematical models describing production rate decline versus time (exponential, hyperbolic, harmonic, and other). The following "decline curves" from production wells are commonly used in DCA (see Figure 2.16):

- production rate vs. time;
- production rate vs. cumulative oil production;
- water cut vs. cumulative oil production;
- GOR vs. cumulative production;
- percentage oil production vs. cumulative oil production;
- $(\overline{p}/z)$ vs. cumulative gas production.

2.6.3 Simplified forecasting (engineering) methods

As was mentioned above, it is rather common practice in reservoir engineering to express the oil recovery factor as a product of *volumetric* (macroscopic) *sweep efficiency* E_S and *displacement efficiency* (microscopic sweep efficiency, as it is often called) E_D:

$$\text{ORF} = E_S E_D \tag{2.3}$$

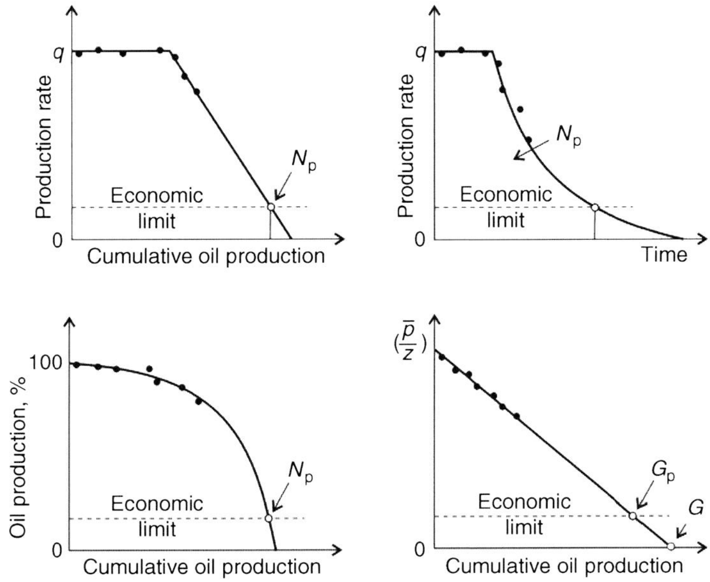

Figure 2.16: Different ways of data representation for a decline curve analysis (Zolotukhin & Ursin, 2000).

In cases when the volumetric sweep is not known or its direct determination is difficult, its decomposition on *area* E_A and *vertical* E_V *sweep* efficiencies can be done, which results in the following expression for the oil recovery factor:

$$\mathrm{ORF} = E_A E_V E_D \tag{2.4}$$

In cases when data on displacement efficiency are not available, E_D is replaced by its ultimate (final) value, which is reached at 100% water cut, that is:

$$E_D = \frac{1 - S_{win} - S_{ores}}{1 - S_{win}} \tag{2.5}$$

where S_{win} is initial water saturation and S_{ores} the residual oil saturation after water flood.

Figure 2.17 highlights typical ranges reported for waterflooding sweep efficiencies and recovery factors in the literature. The recovery factors stated are not derived multiplying the sweep efficiencies given in the table, since the high and low factors usually do not all occur in a given reservoir at the same time.

Instead of using the values given in Figure 2.17, the *volumetric sweep efficiency* E_S can be approximately described as a function of the inter-well distance L and mobility ratio M in the following way:

$$E_S = L^{-b[M/(M+1)]^d} \tag{2.6}$$

where b and d are constants depending on reservoir properties (heterogeneity, permeability contrast, etc.) that can be determined by running few times a simulation model for the actual

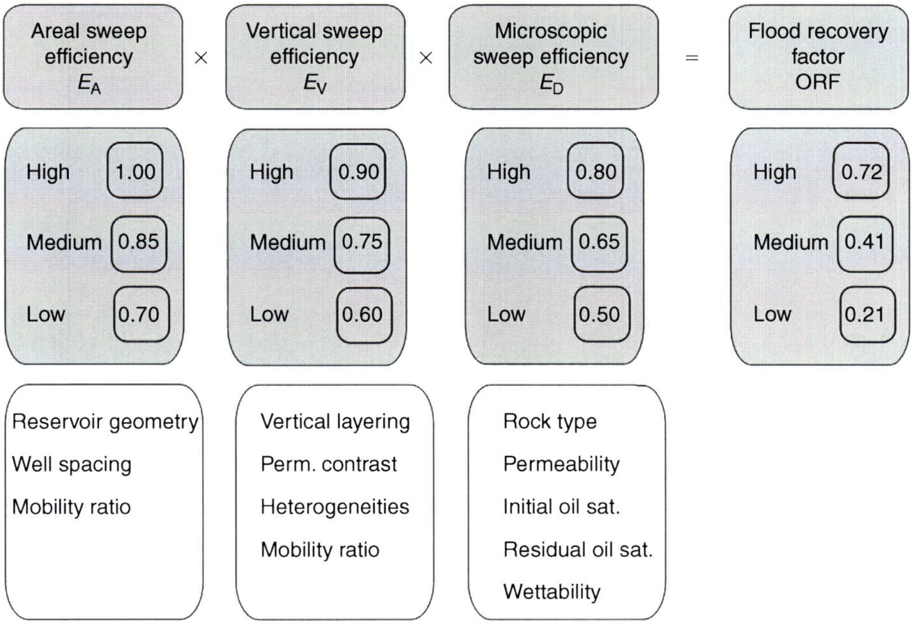

Figure 2.17: Typical ranges for macroscopic and microscopic sweep efficiencies and the oil recovery factor for waterflooding.

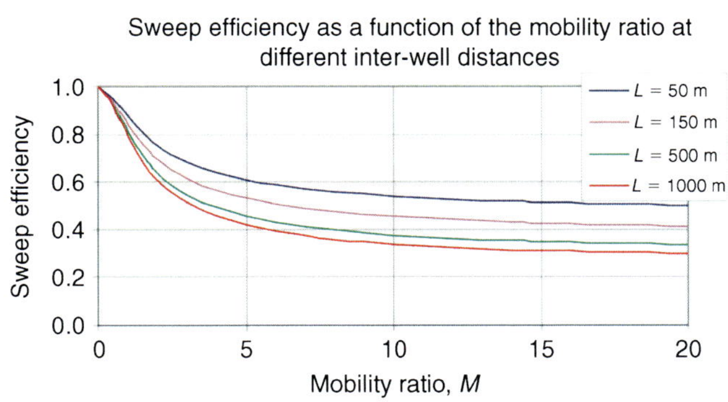

Figure 2.18: Macroscopic sweep efficiency as a function of the mobility ratio at different inter-well distances. Illustration of results obtained by stochastic reservoir simulation (Zolotukhin, 2000).

reservoir. This relationship was obtained by using results of stochastic numerical reservoir simulation model (Zolotukhin, 2000).

Figure 2.18 shows dependency of the microscopic sweep efficiency E_S on the mobility ratio M at different inter-well distances L. Results of several runs of stochastic reservoir simulation enables identification of the values of constants b and d.

2.7 Reservoir risk and uncertainty

Usually a decision to fully develop a field has to be taken without as much information on the prevailing reservoir conditions as would be needed to identify the most cost-efficient development solution. Recognizing that any description and prediction of reservoir performance is uncertain, it is necessary to quantify the uncertainties and identify the parameters most significantly contributing to the uncertainties.

There are several causes of uncertainty, resulting from incomplete or poor data or from inadequate data acquisition, processing, interpretation, integration, and, in difficult cases, data associated with very complex reservoirs. All these uncertainties combined with the geological assumptions finally lead to a complex reservoir modeling which enables us to take into account the different physical phenomena that occur in the reservoir, though highly populated by a number of uncertain parameters.

The reservoir model is a guide to and background for important decisions like estimation of available reserves, hydrocarbons in place, production profiles, economic evaluation, etc. If the base model is built from uncertain parameters, then obviously all these outputs are also uncertain. The field development and succeeding production therefore must reflect the existence of major reservoir uncertainties, which can impact well locations and numbers, completion designs, operations, handling capacities, and economics.

Uncertainty related to reservoir description and performance prevails throughout the life of any field, and is at its highest early in a field development, but diminishes with time. The *quantification* of uncertainties aids in describing the resulting economic span and helps decision makers understand the risk associated with a decision. It can further be used to optimize investments in a flexible design in order to meet future possible deviations from the base assumptions. The *identification* of the most uncertain parameters aids in establishing a rational reservoir monitoring and data acquisition program, to reduce the uncertainties in a cost-efficient manner.

2.7.1 Reserve uncertainty in the exploration phase

At early stages of a prospect/field evaluation, an estimate of the resources and recoverable reserves is associated with a large uncertainty due to low resolution of seismic data and (laterally) sparse log/core data, insufficient for a detailed description of the productive formation. Reservoir thickness, net-to-gross ratio, sealing capacity, source rocks, and potential conduits are largely derived from the well data, as are reservoir and fluid attributes. As a result, the volumetric estimate of resources and reserves is approximate since all the input parameters, like the rock volume, porosity, net-to-gross ratio, initial oil/gas/water saturation, etc., are known approximately, thus reflecting only our incomplete knowledge.

There are several ways to evaluate the uncertainty involved in the volumetric estimate of resources: it could be done based on a deterministic approach and a sensitivity analysis, by probabilistic methods (Monte Carlo simulation) or by fuzzy methods (fuzzy arithmetic and interval calculus) as have been covered in the previous chapter. At early stages of a prospect or field appraisal, probabilistic and fuzzy methods are more appropriate techniques enabling us to appreciate the uncertainty of the reservoir/fluid parameters in a most natural way. Usually results of these evaluations can be presented in the form of the cumulative probability (possibility) distribution functions where the amount of resource or reserves is associated with the probability (possibility in case of fuzzy calculus) of having such a volume. At later stages, more information on the reservoir becomes available and, thus, the deterministic estimate can be applied.

The following notations are in common use:

- An estimate with 90% probability is regarded as *proven* resources/reserves.
- An estimate with 50% probability is regarded as *proven* + *probable* resources/reserves.
- An estimate with 10% probability is regarded as *proven* + *probable* + *possible* resources/reserves.

2.7.2 Uncertainty in planning for production

2.7.2.1 Uncertainty in production profile

Expected production profiles can be generated through the construction of simulation models or by assessing single well performances based on the reservoir description and using a spreadsheet approach.

A production profile (forecast) is then generated based on proposed drilling sites, assessment of drilling and completion times, well spacing, well types, completion, and productivity. The production forecast defines the requirements for processing capacities and storage as well as pipeline infrastructure. In a full cycle development planning, operational elements related to drilling, production, maintenance, and data acquisitions are also assessed. A typical output of such a prediction is shown in Figure 2.19.

During planning for the production phase, the uncertainty involved is fluid flow and productivity parameters such as relative permeability, PVT, as well as number of wells and well architecture. The additional data on geology concepts, stratigraphic, and petrophysical properties we get so far from exploration/appraisal wells are based on single-point wells and from these data we extrapolate for the entire reservoir. The decisions based on these data and the forecast of the production profile are therefore uncertain.

The well productivity at this stage can be estimated by using a simple Dupuit's equation (this equation is often and erroneously called Darcy's equation; however, it was Dupuit, a French engineer, who obtained this equation in 1863 based on Darcy's experiments):

$$q = \frac{2\pi kh}{\mu B} \frac{p_e - p_{wf}}{\ln(r_e / r_w) + S} \tag{2.7}$$

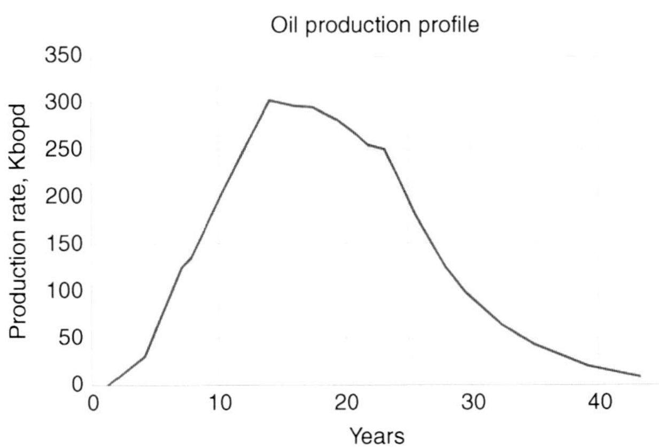

Figure 2.19: A typical production profile from simulation.

where k is the reservoir permeability, h the reservoir thickness, μ the viscosity of oil, B the oil formation volume factor, p_e the pressure at the drainage area boundary, p_{wf} the flowing well (bottom hole) pressure, r_e the radius of the drainage area, r_w the wellbore radius, and S the skin factor (by using a skin effect concept, a reservoir engineer can evaluate productivity (injectivity in case of injection well) of various types of well like horizontal, slant, partially completed, fractured, etc., by assigning appropriate (approximate) value of skin factor S).

Note that almost all the input parameters in the above equation are known approximately.

Here the uncertainty involved in the productivity evaluation can be estimated by using the same approaches as in case of resources and reserves estimate. As follows from Dupuit's formula (2.7), the well productivity depends on reservoir and fluid properties; thus, the uncertainty in the reservoir description directly influences the uncertainty in the well productivity evaluation. Usually, a common practice in most of the companies is to limit the number of possible outcomes of probabilistic resource estimates by those corresponding to three levels of certainty (e.g., P_{90}, P_{50}, and P_{10}).

Each possible outcome of a resource estimate can be referred to as a *scenario* (which, in contrast with a *concept*, does not depend on our decision). Considering three such scenarios (note that these three scenarios correspond to a single field development concept), an evaluation of well productivity can be done for each of these scenarios with numerical values of reservoir and fluid parameters corresponding to the level of certainty specified in the selected scenario. A field development concept based on possible outcomes should be adequate for all three scenarios. In reality, no development concept can be equally good for all possible outcomes. It means that selected concept should be *flexible* enough to allow necessary adjustments for correcting and optimizing reservoir performance at minimum cost.

Since the well productivity depends on reservoir properties, the field development concept should reflect the differences in reservoir properties "assigned" to each possible outcome (scenario) in terms of number of wells, their completion design, and location. This means that, in principle, the number of wells, their completion, and location (the *drainage strategy*) could be different for each possible outcome. Again, considering three possible drainage strategies for each of the scenarios selected earlier, it will provide nine different scenarios to be considered at this stage. This can be framed as a decision tree problem with growing dimension.

The more branches are involved in a decision tree, the more possible outcomes should be considered. If, in our field development design, we consider three branches, like reservoir, well system, and surface facilities, then, in general, for each conceptual design we should consider $3^3 = 27$ different outcomes! It requires professionalism and experience to select a conceptual design that is flexible enough to account for all possible outcomes while allowing for necessary adjustments at minimum cost.

Figure 2.20 illustrates the uncertainty in a production performance forecast showing production profiles corresponding to three different outcomes of a resource assessment.

The production forecast gives a basis for estimating recoverable reserves for the field (Figure 2.21) and the plateau production volume, normally restricted by estimates of yearly offtake of reserves and well potentials. These yearly offtake estimates can vary within 1–20% of reserves. This value has to be estimated for each field based on ultimate recovery, sensitivity to the offtake rate, and well potentials in the field.

For example, if the limiting offtake rate is 10% per year, then the estimated peak production rate q_{max} is:

$$q_{max} = (\text{Reserves})\left(\frac{0.10}{365}\right)\text{bopd} \tag{2.8}$$

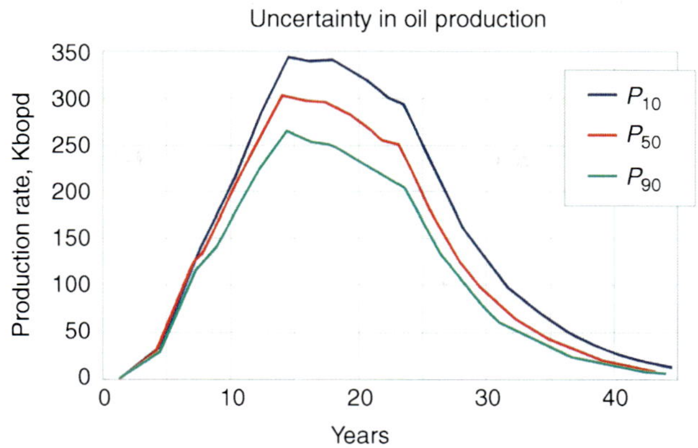

Figure 2.20: Uncertainty in production performance forecast (Zolotukhin, 2000).

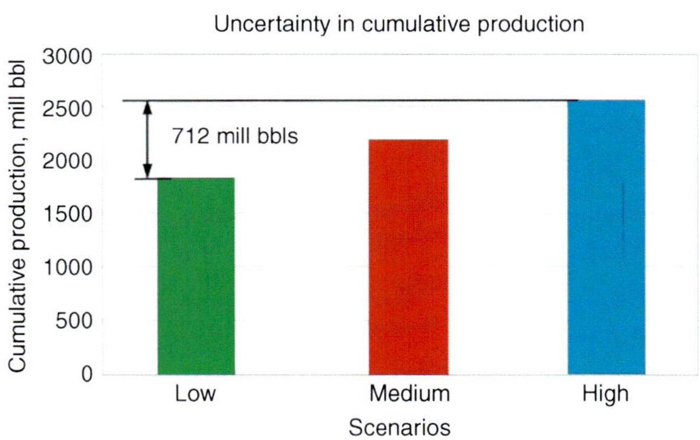

Figure 2.21: Uncertainty in cumulative oil production. Cumulative production values correspond to production profiles shown in Figure 2.20.

The plateau period may last 3–5 years, which indicates that the total production life of field, technically seen, could be as long as 15–20 years, and in some cases longer.

One thing is certain about the production forecast at this stage: *it is not correct!* But there is nothing better available at this stage of the development. The profile has an upside and a downside, and what is really needed are risked values of the production level and of the recovery, as illustrated in Figures 2.20 and 2.21. This is possible by running fuzzy or probabilistic simulation models. Figure 2.22 shows the results of reservoir production forecasting by stochastic simulation.

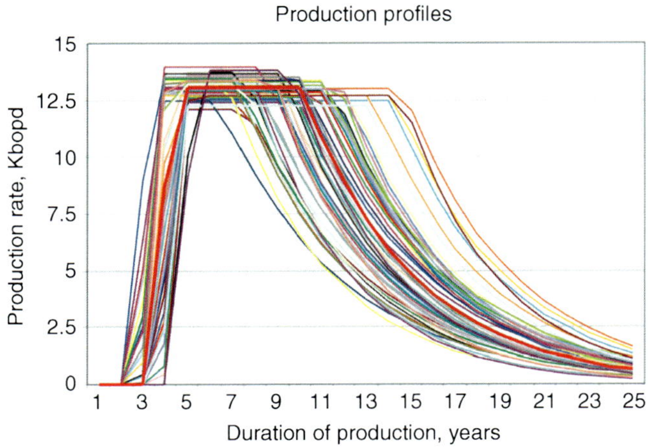

Figure 2.22: Results of reservoir production forecasting obtained by stochastic (Monte Carlo) simulation.

2.7.2.2 The wellstream composition

In addition to planning for a certain production profile, or range of production profiles, it is also necessary to clarify as much as possible the composition of what will come out of the wells. This has important implications for processing facilities and offtake systems. Issues of particular concern include the following:

- associated gas production;
- quantities of natural gas liquids (propane, butane, and light condensate);
- sour (corrosive) gas components in the wellstream (H_2S and CO_2);
- flow properties and other physical properties of the crude oil;
- quantities of water.

The implications of the above issues are addressed in later chapters.

2.7.3 Uncertainty during field operation

During the field operations, the information on the reservoir properties increases, in particular if a continuous field measurement program is undertaken. Based on the available production data and field measurements, which may be accompanied by updated seismic surveys (4D seismic), it can be relevant to carry out *numerical simulations* of the production performance. Such simulations represent an excellent way to select the further production strategy, in particular with regards to injection needs, the optimal extraction of fluids, and the need for new wells. In this respect, the uncertainty regarding the field production is reduced.

2.8 Reservoir management

Reservoir management aims to maximize the economic value of a reservoir by optimizing recovery of hydrocarbons and costs.

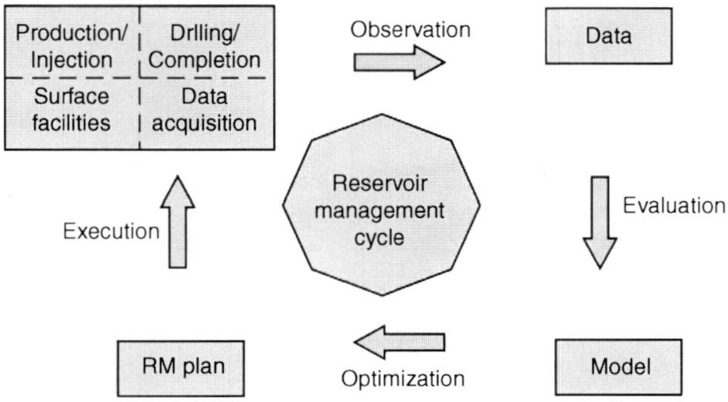

Figure 2.23: Reservoir management cycle.

The emphasis and focus of the reservoir management effort varies in different phases of a field development. Reservoir management is a continuous and cyclic process, as indicated in Figure 2.23.

The arrows in Figure 2.23 indicate *processes*, the squares *products* in the reservoir management cycle. All the processes and products have varying emphasis and degrees of definition in different phases of reservoir assessment and development, but they are usually all present.

Core elements of reservoir management are the perception of reservoir characteristics and production properties, the selection of production strategy (depletion and injection) and reservoir modeling, economic optimization of well types, number of wells, completion strategy, site locations, surface capacities, and a reservoir management execution plan. Other aspects are execution of plans, data acquisition and monitoring of results, updating of model, and repeating the cycle. The reservoir description in terms of structural layout, hydrocarbon distribution and volumes, reservoir quality, etc., is the main contributing element for determination of an optimal production strategy (including injection where relevant), well numbers and types, surface facility layout, and capacities.

The reservoir management cycle therefore represents, in early phases of a field development, the main design basis to adopt surface facilities and drilling to the prevailing perception of the reservoir. In a production phase reservoir, management serves to optimize drilling, production capacities, and operations to a revised perception of reservoir description and performances.

2.8.1 Reservoir management in a prospect evaluation phase

There is usually no well information available for the structure in question in the prospect evaluation phase. Data directly related to the prospect are generally 2D or 3D seismic. Based on these data and discovery and production experiences made in the area, a reservoir description assessment is made, forming the basis for the establishment of a reservoir development strategy and reserve evaluation.

The exploration phase is basically a data acquisition phase enabling a field development decision with acceptable risk associated. In either case, the reservoir management input is quite similar: reservoir description, reservoir development strategy, number of wells and their types and completions, processing capacities, production profiles (oil, gas, and water), and reservoir monitoring and data acquisition plan.

Table 2.1: Cash flow, uncertainty, and IOR impact in the production phases of a field.

	Build-up phase	Plateau phase	Decline phase
Cash flow	Increasing	High	Decreasing
Reservoir uncertainty	High	Medium	Decreasing
Impact of IOR implementation on reserves	High	High to medium	Decreasing

2.8.2 Reservoir management during the production phase

The aim of reservoir management during production is to change or optimize the reservoir development strategy, drilling and completion program, surface capacities, and operations by updating the reservoir description based on reservoir monitoring and data acquisition.

Any change with associated increase in cost has to be justified economically. The production phase typically can be divided in three parts, *the build-up, plateau, and decline phases.* The aspects of reservoir management changes with the phases are shown in Table 2.1.

Based on Table 2.1, it can be justified that focusing and front-loading reservoir management activities is a suitable strategy for optimizing production. In the first two phases, both reservoir uncertainty and impact on production are high at the same time as cash flow is increasing. Cash flow, reservoir uncertainties, and production impact are lower in the decline phase, suggesting a lower level of reservoir management activities.

The other aspect of reservoir management is to evaluate and substantiate major changes in production strategy (IOR) in terms of injection, well types and completion, and the associated modifications in surface facilities. These opportunities generally appear as a result of revised reservoir description and/or understanding of reservoir dynamics, revised availability, and cost of injection fluid or development of new technology appropriate for the development.

2.9 Summary

Reservoir and production management is a core business in upstream petroleum industry. Reservoir characterization, fluid flow, and reservoir management are the key parameters that need a full attention of the reservoir engineers. These parameters, unlike other parameters analyzed in the complete process of field development, are always evaluated in the development phase for the entire lifetime of the field to ensure maximum optimization.

References

Amyx, J.W. Jr., Bass, D.M. & Whiting, R.L. 1960. *Petroleum Reservoir Engineering.* McGraw-Hill, New York.

Archer, J.S. & Wall, P.G. 1991. *Petroleum Engineering, Principles and Practice.* Graham and Trotman, London.

Bedrikovetsky, P.G. 1993. *Mathematical Theory of Oil and Gas Recovery.* Kluwer, Dordrecht.

Dake, L.P., 1978. *Fundamentals of Reservoir Engineering,* 12th ed. Elsevier, Amsterdam.

Frick, T.P. & Zolotukhin, A.B. 1994. *A Mobility Driven Fingering Approach to a Field-scale Simulation of Oil Recovery.* SPE Paper 27017. Proceedings of the III Latin

America and Caribbean Petroleum Engineering Conference, Buenos Aires, Argentina, 26–29 Apr.

Havlena, D. & Odeh, A.S. 1963. The material balance as an equation of a straight line. *Journal of Petroleum Technology,* 15(8) pp. 896–900.

Havlena, D. & Odeh, A.S. 1964. The material balance as an equation of a straight line, Part II – field cases. *Journal of Petroleum Technology,* 16(7) pp. 815–822.

Klinkenberg, L.J. 1941. The permeability of porous media to liquids and gases. In: *Drilling and Production Practices.* American Petroleum Institute, pp. 200–213.

Larue, D.K. & Yue, Y. 2003. How stratigraphy influences oil recovery: a comparative reservoir database study concentrating on deepwater reservoirs. *The Leading Edge,* 22(4) pp. 332–339.

Zolotukhin, A.B. 2000. Dynamic Fractal Models for Two-Phase Flow through Porous Media. In: Dmitrievsky, A. & Panfilov, M. (Eds.), *Porous Media, Physics, Models, Simulation.* World Scientific Publishing Co. Pte. Ltd., Singapore.

Zolotukhin, A.B. & Ursin, J.-R. 2000. *Introduction to Petroleum Reservoir Engineering.* Høyskoleforlaget, Kristiansand, Norway.

Drilling, well design, and well completion

Drilling is the penetration of rock formations using mechanical equipment. In the oil industry, the term "drilling" often refers to the wider activity of constructing a well. Wells are constructed for the purposes of exploration, production, or injection, which require different methods of well planning and completion. This chapter reviews the objectives of drilling, drilling facilities, and rig types available. Criteria for well design, well completion methods, and well testing are also discussed. Health, safety, and environmental (HSE) aspects of drilling are briefly reviewed.

3.1 Types of wells

Although a petroleum production history accounts for more than 150 years of non-stop technology development and improving tools and equipment, a well remains the only conduit connecting a reservoir with the surface. Selecting the optimal drainage strategy, for example, total number of wells, their type, and placement within a reservoir, is one of the most challenging tasks in petroleum reservoir engineering.

There are different types of wells that are used for the development of oil and gas fields:

- exploration wells;
- appraisal and delineating wells;
- production wells;
- injection wells.

Often the reservoir energy is not high enough to maintain production at the desired level. In such cases, injection wells may be used in order to keep the reservoir pressure sufficiently high, thus securing a more stable production.

Depending on the numbers and locations of injection and production wells, different well placement systems can be identified. For example, injection wells can be placed between production wells in a certain repeated fashion called a *well pattern system*. Among the most common are:

- staggered line drive;
- inverted five-, seven-, and nine-spot pattern systems;
- peripheral system.

Each pattern system can include injectors and producers that can be vertical, slant, and/or horizontal. Multilateral wells can also be used in a repeated pattern system. The ratio of injectors to producers indicates the intensity of the system. Intensive systems have more injection wells, with the ratio between injectors and producers close to 1. Such systems are commonly used for developing low-permeable, heterogeneous reservoirs and/or formations containing heavy oils. They enable injection and production rates at substantially high level.

For offshore fields, the total number of wells is often lower than that for onshore fields because of the high cost of offshore drilling operations.

3.2 Drilling

The objective of constructing wells is to create a conduit to one or several reservoirs at some depth below surface. This conduit will be used to evaluate the amount of hydrocarbons in the exploration and appraisal phases, and for production or injection in the exploitation phase.

For many years after the beginning of the oil industry, drilling took place onshore only, but now a large portion of global drilling activity is offshore. Offshore drilling is generally more expensive than onshore drilling: the day rates for renting and operating an offshore drilling rig have historically been three to four times as large as the corresponding onshore rig rates. Offshore drilling also requires more challenging logistics and higher subsequent operational costs.

3.2.1 Objectives of drilling

3.2.1.1 Drilling for exploration and appraisal

On the basis of the seismic interpretation and the regional geological context, a geological prognosis is derived with inputs from the geological and geophysical disciplines. An exploration drilling program is elaborated if the geological prognosis indicates that interesting structures exist.

Exploration wells are wells drilled purely for exploratory purposes in a new area, that is, for obtaining information. The data to be obtained include rock cuttings, cores, drill stem tests, and wireline logs, as well as more sophisticated measurements like borehole resistivity images, obtained through the formation micro-scanner.

Appraisal wells are wells drilled to assess characteristics, such as flow rate, of a proven hydrocarbon accumulation.

The *exploration drilling program* specifies how the target will be reached, the target being a prospective hydrocarbon reservoir. The program contains detailed plans for the location of the drill path, expected formation characteristics, the expected amount of time required for drilling, etc. In the exploration phase, wells are often plugged and abandoned after the necessary data have been gathered. Appraisal wells may be temporarily suspended and kept, in order to be used as producers sometime in the future.

Data collection, analysis, and model evaluation in the exploration phase is an iterative process as indicated in Figure 3.1. After a field discovery, subsurface sampling provides more data with which to refine a reservoir model. Though models are always approximations, they help to improve the understanding of a field as discussed in the previous chapter. Sometimes the data obtained may change the reservoir model to such an extent that it can result in the termination of operations and even the abandonment of the field.

3.2.1.2 Deciding on exploration or appraisal drilling

Exploration and appraisal drilling are done for the purpose of obtaining information. Such information is a necessity for any field development to take place. The cost of drilling must therefore be justified by the value of the information that can be obtained. The information value of an exploration well derives from the potential value of a discovery that may be made, considering

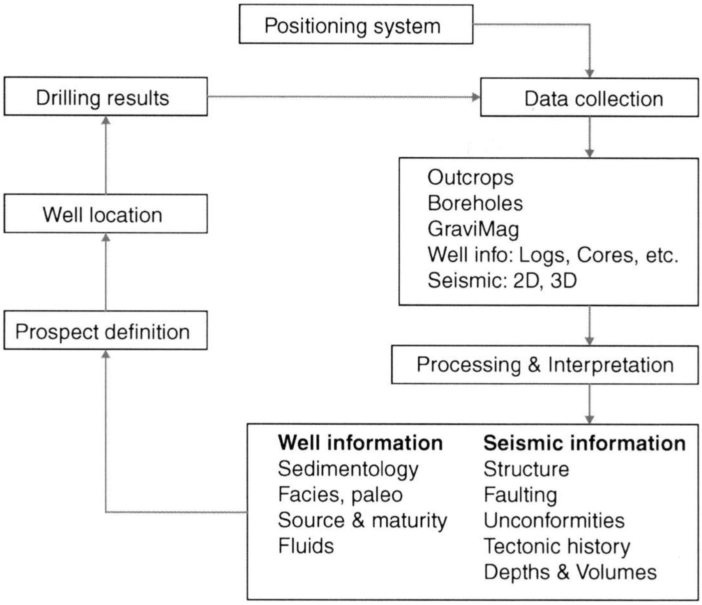

Figure 3.1: Data collection and model evaluation.

also the probability of making or not making a discovery. The information value of an appraisal well derives from how the information from it will contribute to reaching good decisions on developing the field. Reducing the uncertainty in reserve estimates is often an important consideration for appraisal drilling.

In the context of exploration and appraisal drilling, the term *risk* is often used as the probability of obtaining an unwanted result, such as not finding any hydrocarbons which will be worth producing.

A well to be drilled has one or several drilling targets, that is, specific locations in the underground which should be reached by the well. The selection of precise drilling targets will greatly influence the value of information which will be obtained from drilling. Based on prior available information, such as seismics and knowledge of similar drilled structures elsewhere, it may be possible to estimate expected values for several parameters which will be important criteria for the selection of drilling targets, in particular the following:

- the probability of hydrocarbons being present in the prospect;
- the prospect size, determined by its area, vertical roll-over, integrity of closure, depth to crest, and hydrocarbon column;
- reservoir characteristics, including net to gross (N/G) ratio, porosity, permeability, and reservoir pressure and temperature;
- the presence (saturation) of various hydrocarbon fluids, that is, oil, gas, and natural gas liquids (NGLs).

Also based on prior available information on the geology in the area, a well prognosis should be established prior to drilling, providing expectations for the total depth of drilling, lithology, pressure, and prediction of oil and/or gas.

3.2.1.3 Drilling for development: production wells

Production wells are wells drilled primarily for producing oil or gas, once the producing structure and characteristics are established.

In the case of a development well, all the data that serve to define the drilling target will be known, but additional problems associated with well completion, stimulation, and activation and proximity to other wells must be considered. During a field development, production wells, water injection wells, gas injectors, and disposal wells are normally drilled, although all are not necessarily required in one particular field development. Wells may be re-entered and altered or repaired through the lifetime of the field.

3.2.2 Drilling facilities

3.2.2.1 Drilling rigs

A rig is a machine used to drill a wellbore (see Figure 3.2 for an example). In onshore operations, the rig comes with most auxiliary facilities required to carry out drilling operations, not including living quarters. Major components of the rig include the mud tanks, the mud pumps, the derrick or mast, the draw works, the rotary table or top drive, the drill string, the power generation equipment, and auxiliary equipment. Offshore, the rig includes the same components as onshore and in addition there are the facilities of the vessel or drilling platform on which the rig is placed. The rig is sometimes referred to as the drilling package, particularly offshore.

The main technical systems of oil well drilling rigs are:

- power generation;
- drill string;
- hoisting system;

Figure 3.2: A land rig in operation in Kazakhstan. Photo: courtesy of Parker Drilling Company.

- rotation system;
- circulation system;
- pressure control system.

3.2.2.2 Power generation

Because of large variations in power requirements and their mobile nature, rigs are self-contained for power and electric distribution system. The power generators are often diesel driven, and they use silicon-controlled rectifiers (SCR) to convert AC power to DC.

3.2.2.3 Drill string

The basic process of drilling relies on energy being used to break up the formation and create a hole. The energy is transmitted to the formation through the use of a drill string, through rotation, weight on bit, and hydraulic forces. The drill string is a hollow, high-quality steel pipe (tubulars), capable of transmitting torque to the drill bit located at the end of the drill string. Simultaneously, fluid is pumped down into the drill string, and is jetted through the drill bit at high pressure and force, removing all the cuttings.

Different types of drill bits are used corresponding to different rocks which need to be penetrated. The most common type is a drill bit with three rotating cones where the drill bits cut through the softer rock minerals. These rotating cones get their motion from the rotation of the drill pipe. Very hard rock types need to have diamond drill bits where artificial diamonds are fitted into many small cavities. The drill bit gets worn out in the drilling process. The frequency with which the drill bit needs to be changed is determined by the strength of the rocks to be penetrated.

3.2.2.4 Hoisting system

The hoisting system consists of two main components: the *drilling derrick* or *mast* and the *draw works*. The derrick has the main function of supporting the loads of pulling and running tubulars in and out of the well. The draw works provides force to lift the tubulars.

A new concept in hoisting equipment for drilling rigs has been introduced in the market. This relies on hydraulic cylinders for hoisting and lowering the drill string and other tubulars in and out of the well. This takes away the need for a high-strength derrick, as the loads are taken up by the hydraulic cylinders. No draw works are required. In comparison with conventional drilling rigs, this concept reduces the total weight of the drilling package, and also reduces the safety risks in the drilling process as there are less moving parts (Figure 3.3).

3.2.2.5 Rotation system

There are two types of rotation systems in drilling operations, namely rotary table and top drive systems (TDS). The rotary table is located in the drill floor, and has a motor which drives the table around through a clutch and gear arrangement. The rotary table motion is transferred to the drill string through a bushing locked into the rotary table (see Figure 3.4). The top part of the drill string is the kelly, a square or hexagonal pipe, which fits through the bushing and acts as a direct drive for the drill string.

The widely used TDS has over the past 20 years made the rotary table for rotation redundant on modern rigs. The top drive is hung underneath the traveling block, where it is guided by rails to arrest any torque in the unit. The top drives have an electric or hydraulic motor, gear arrangement, and screws directly into a full stand (three joints) of drill pipe. It is considered to be more efficient than the normal rotary table, particularly when drilling deviated wells.

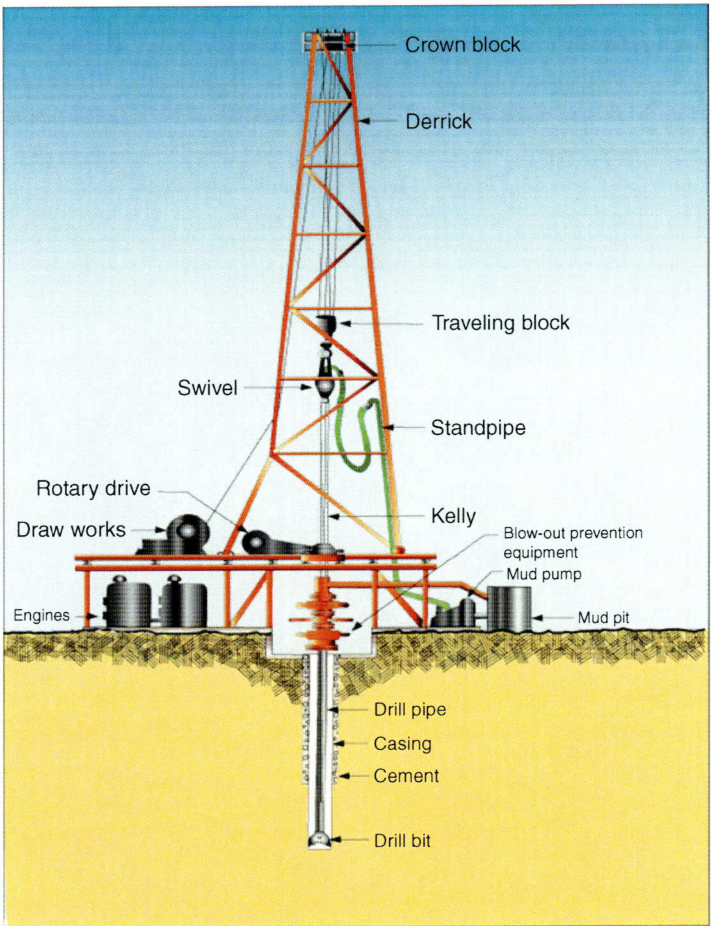

Figure 3.3: A sketch of a drilling rig with rotary table and its components in operation. Illustration: Department of Conservation, State of California.

3.2.2.6 Circulation system

In order to provide the hydraulic power required to help break up the formation and carry cuttings to surface, a circulating system is used. The main items in this system are the circulating fluid, more commonly known as *drilling mud*, mud pits, high-pressure pumps, and drilling fluid mixing/treatment equipment (Figure 3.5). The mud pits provide storage for different types of fluids when required. They act as a buffer for the drilling fluid volume, and give the possibility to mix and add chemicals to the drilling fluid during the operation.

Cuttings generated in the drilling process are removed from the fluid on its return from the well by passing the fluid through the solids control system. This consists of vibrating screens called shale shakers and centrifuges (hydrocyclones) to remove finer particles from the mud. The mud is passed back to the storage tanks, and pumped back into the well. Another part of the circulating system is the cementing unit, which is a high-pressure cement mixer and pump, primarily used to place cement around casings in the well. Other uses include the function as a backup for the circulating system and for pressure testing equipment.

Figure 3.4: Drilling operations on a floating rig with a rotary table. Photo: Statoil/Ø. Hagen.

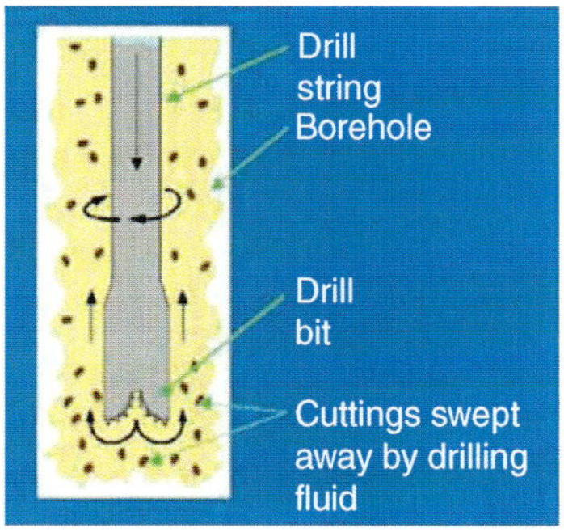

Figure 3.5: A sketch showing drilling mud pumped down into the hole and returning to the surface laden with cuttings.

3.2.2.7 Pressure control system

The hydrostatic pressure exerted by the drilling fluid is the primary barrier for preventing hydrocarbons flowing to surface during a drilling operation. The pressure control system is the secondary barrier. The primary barrier can fail due to various reasons, leading to losses of drilling fluid to the formation. Such loss reduces the hydrostatic column and thereby hydrostatic pressure exerted by the drilling fluid. This will cause the formation pressure to exceed the hydrostatic pressure of the drilling fluid, thereby causing hydrocarbons to flow into the wellbore, referred to as a "kick." A kick situation can lead to an unstable well control situation, referred to as a blow-out.

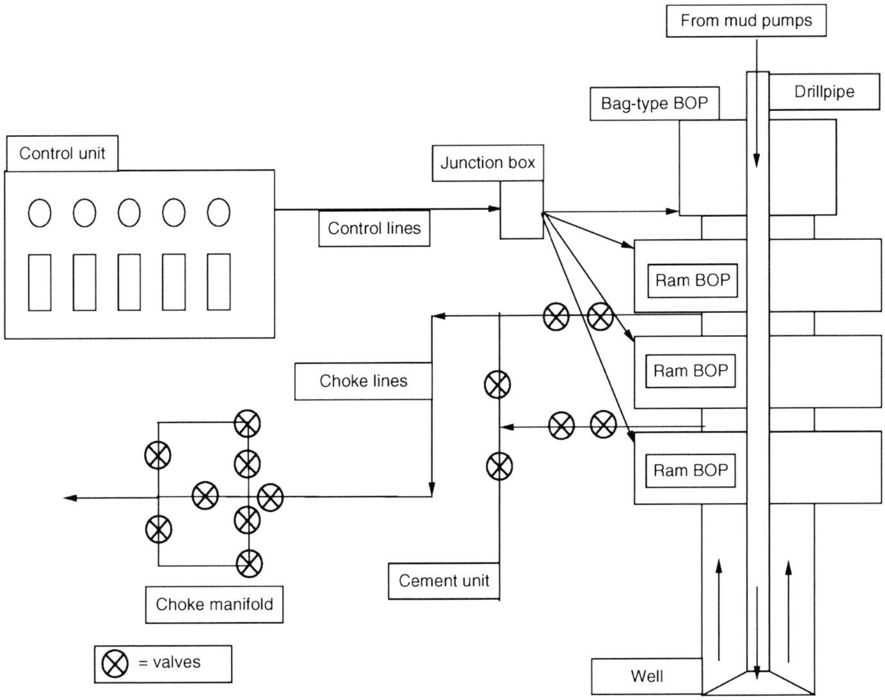

Figure 3.6: A schematic sketch of a pressure control system.

In all these cases, the pressure control system is employed to ensure safe operations. This system has three main components: the blow-out preventers (BOP), the choke and kill lines with choke manifold, and the control system (see Figure 3.6). The BOP are placed directly underneath the drill floor for platform, jack-up, and land operations or, in the case of semi-submersible rigs and drillships, on the seabed.

There are several side outlets on a BOP stack. They allow mud to be circulated around the well to regain control where hydrocarbons or high-pressure water have entered. When circulating out an influx in the well, the drilling fluid is pumped down through the drill string, up the annulus, and through a pressure-reducing choke valve. Control of the choke opening allows any such influx to be circulated out while maintaining control of the well.

3.2.3 Drilling mud

Drilling muds are mixtures of natural clays and heavy minerals (usually barium sulfate), and small amounts of several other chemicals suspended in water or an organic liquid to influence the viscosity of the fluid.

There are three types of drilling mud:

- oil-based drilling mud (OBM);
- water-based drilling mud (WBM);
- synthetic-based drilling mud (SBM).

3.2.3.1 Oil-based drilling mud

Oil-based muds excel in drilling performance through elimination of water in the external phase and the positive contributions of the oil which is non-polar and lubricious. OBM is characterized by providing the borehole stability, cuttings stability, efficient solids removal, superior lubrication, and greater rate of penetration. Despite their broad applicability, high performance, and capacity for reuse, OBM carry relatively high unit costs. They are more difficult to mix, and prone to lost circulation. They are also less environmentally acceptable than water-based muds especially if untreated oily cuttings are discharged.

The environmental concerns make the use of oil-based muds increasingly more difficult, costly, and risky since special authorization must be obtained, and expensive cuttings-disposal procedures need to be implemented.

3.2.3.2 Water-based drilling mud

Water-based mud and cuttings are less toxic and therefore are environmentally friendlier than oil-based muds. However, WBM may not work well, particularly when drilling a very deep or deviated well. It also has undesirable characteristics including the reaction with formation clays and shales, hydration, plasticization and/or disintegration of cuttings, loss of wellbore support through pore pressure transmission, and the tendency to dissolve salt and corrode metals.

3.2.3.3 Synthetic-based drilling mud

SBM was developed to combine the favorable drilling properties of OBM with the low toxicity of WBM. The liquid phase of SBM is a synthetic organic chemical, usually an ester or olefin. SBMs have been in use since the 1900s for drilling, particularly in the North Sea and Gulf of Mexico. Only the cuttings produced with the use of SBM are discharged to the ocean; the drilling mud itself is separated from the cuttings and recycled or regenerated. Studies are however carried out to determine the long-term effects of cuttings discharge on the ocean environment. The results of these studies will be used to develop strategies for minimizing environmental impacts of cuttings discharge to the ocean.

3.2.4 Types of drilling rigs for different working environments

As petroleum exploration and production has moved into new areas with new complexities and in different working environments, drilling rigs have been installed on different types of fixed and floating foundations in order to be available at desired locations. For use on water, there are drill barges, submersible rigs, jack-up rigs, semi-submersible rigs, drillships, and platform rigs. Several of these different types are illustrated in the subsequent figures.

3.2.4.1 Land rigs

Small land rigs are moved between locations by automotive power as a truck-mounted unit. Larger type land rigs are dismantled and moved in sections by trucks and trailers. There are some smaller helicopter-transportable units for special applications.

3.2.4.2 Drilling barges

Drilling barges are large, floating, bottom-supported units which are towed (moved) into the proposed location in floating condition. They are then counterbalanced to rest on the bottom prior to commencing the drilling operations. They are often used for inland shallow waters like in swamps, lakes, or other shallow-water areas. They are suitable for calm and shallow waters, and not suited to withstand the water movements experienced in open seas.

3.2.4.3 Submersible rigs

Submersible rigs, suitable for shallow water, also come in contact with the ocean or lake floor. These rigs consist of a platform with two hulls positioned on top of one another. The upper hull contains the living quarters for the crew, as well as the actual drilling platform. The lower hull works much like the outer hull in a submarine vessel: when the platform is moved from one place to another, the lower hull is filled with air-making the entire rig buoyant. When the rig is positioned over the drill site, the air is let out of the lower hull, and the rig submerses to the sea or lake floor. This type of rig has the advantage of mobility in the water; however, once again its use is limited to shallow-water areas.

3.2.4.4 Jack-up rigs

A jack-up is a mobile drilling rig, which will reach the drilling location by self-propulsion or with the assistance of tug boats (Figure 3.7). It has long leg structures which it lowers into the seabed, raising the rig out of the water. The obvious limitation with this type of installation is the depth of water it can operate in, the maximum being about 500 ft. A jack-up rig may also be used as a tender support vessel (TSV) alongside a wellhead platform, as will be explained further below.

3.2.4.5 Semi-submersible rigs

Semi-submersible rigs, shown in Figure 3.8, are the most common type of offshore drilling rigs, combining the advantages of submersible rigs with the ability to drill in deep water. They work on the same principle as submersible rigs except that semi-submersible rigs are held in place by large anchors, each weighing upwards of 10 tons. These anchors, combined with the submerged portion of the rig, ensure that the platform is stable and safe enough to be used in turbulent offshore waters.

Some of the modern-day units have a dynamic positioning (DP) system which relies on propellers and accurate navigation systems to maintain position. The DP units are often preferred in deep-water areas, where installing anchors is costly and time consuming.

Figure 3.7: A jack-up rig under towage. Photo: Statoil/K.G. Poulsen.

Figure 3.8: A semi-submersible rig. Photo: Statoil/H. Schjølberg.

The semi-submersible rig is used to drill in water depths from some 60 m to 2500 m and beyond, depending on age, type, and specification of the rig. They are moved between drilling locations by towing or are self-propelled.

3.2.4.6 Drillships

Drillships are self-propelled units, generally with a large payload capacity. They can therefore carry large amounts of consumable materials and equipment in order to operate in remote areas where supplying the drilling operation using boats is costly. Drillships have also been widely used for deepwater drilling. As the name suggests, this is a ship shaped as a drilling vessel (see Figure 3.9). It does not require tugboats to tow it to location. Although drillships are not as stable as semi-submersibles, they also drill in very deep waters.

3.2.4.7 Platform rigs

Platform rigs are installed on production platforms and are used to drill and service the production wells (Figure 3.10). The total number of wells drilled from one platform vary depending on well and reservoir performance, but is in general limited upwards to some 40–50 wells. The platform rig is in most cases permanently installed on the platform, but may be moved off and replaced by well servicing equipment when all planned wells for a particular platform have been drilled and completed.

3.2.4.8 Tender support vessels

A TSV is used to support a drilling operation on a platform, and usually has the form of a barge or a semi-submersible. It is mainly used in offshore areas like the Far East or Venezuela where the water depths are limited, wave conditions are mild, and many small platforms can

Figure 3.9: A drillship equipped with dual drilling derrick. Photo: © Seadrill.

Figure 3.10: A jacket platform with a drilling rig. Photo: Statoil/H. Hansen.

be built cheaply. The TSV anchors up alongside a small wellhead platform, and has all the drilling equipment onboard in order to drill a well from the platform while alongside. When the well(s) have been completed, the TSV is moved to the next platform.

3.2.5 Advanced drilling techniques

3.2.5.1 Directional drilling

Directional drilling is a technology that allows an accurate control of the inclination and direction of wellbores at relatively minor additional cost over the vertical wells. Large numbers of directional and horizontal production wells have been drilled since the introduction of this technique. The technique is illustrated in Figure 3.11. The main advantages of directional (deviated) drilling can broadly be summarized as follows:

- It allows better well coverage of a reservoir from one central location. This implies fewer rig moves, less surface area disturbance, and cost savings in completing and producing the wells.
- Deviated or horizontal drilling exposes more length of reservoir in one well.
- For horizontal wells, the production rates are often two to four times higher than those for a vertical well.
- It allows drilling into the reservoir where vertical access is difficult or not possible; for instance, an oilfield under a town, under a lake, or underneath is very difficult to drill formation.
- It allows simple and controlled sidetracking of wellbores.

Such sidetracks may be made due to redefined geological targets, technical problems in the current well, or sidetracks planned in existing wells.

3.2.5.2 Horizontal drilling

The ultimate benefit of deviated drilling is obtained through the use of horizontal wellbores, as illustrated in Figure 3.12. Such wellbores can access long sections of the reservoir, providing increased well deliverability compared to vertical drilled wells and higher ultimate field recovery. In many field development cases, horizontal drilling also allows for a reduced number of wells needed for effective reservoir drainage. This may result in cost and production improvements and the possibility to bypass environmentally sensitive areas.

Horizontal wells cannot effectively be applied in all reservoirs. Multilayered reservoir formations or extensive faulting with varying oil–water or gas–water contacts may result in less

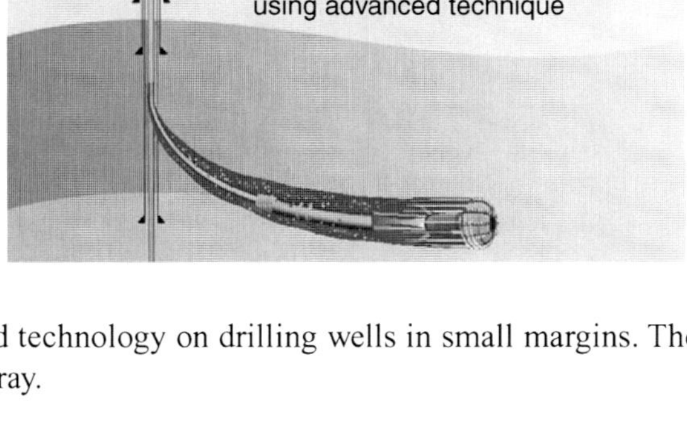

Figure 3.11: Advanced technology on drilling wells in small margins. The reservoir is indicated in light gray.

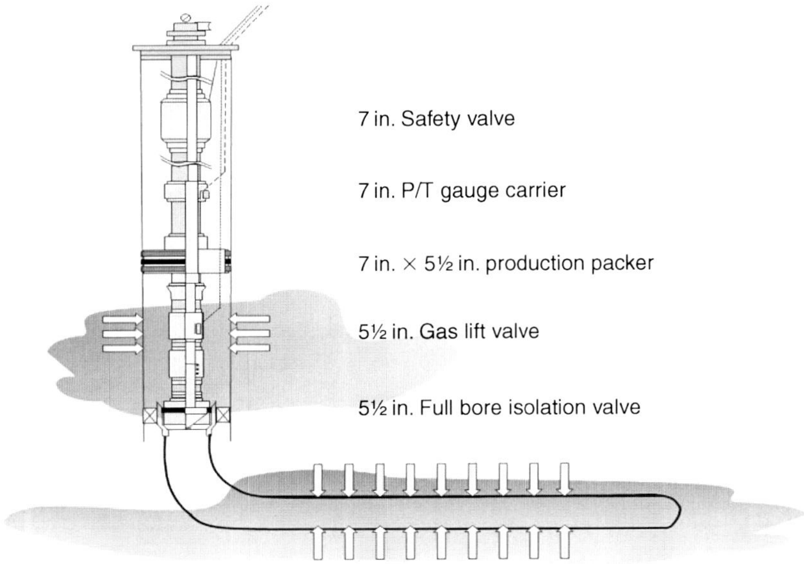

Figure 3.12: Production from a horizontal well. The shaded part indicates an oil reservoir.

coverage with horizontal wellbores than that in the case of a less-deviated well. Reservoirs have different characteristics, the careful study of which should provide the design basis for all drilling- and completion-related work.

3.2.5.3 Multilateral drilling technology

During the 1990s, the technology of drilling branches in the reservoir was developed. This technology enables the drilling of more than two wells (commonly three to six or up to eight production bores) out from a "mother" bore in the production zones of interest. In certain circumstances, the technology has provided access to new reserves and allowed large reductions in overall well costs.

3.2.5.4 Combination well

Figure 3.13 illustrates a well drilled for two purposes: production from a thin oil zone (1) and gas production from a gas cap (2). The horizontal well section is suited to extracting oil from a thin oil layer, which in the illustrated case has a gas cap above and a water zone below. Production from such a well can better be optimized with a low reservoir draw down, to avoid provoking the water to cone or gas to cusp into the wellbore prematurely. After production of the oil zone, gas can be produced through the higher perforations (2), to avoid drilling a new well for the purpose.

3.3 Well design

Before drilling a deep, complicated, and expensive well, thorough planning is required. To ensure that all available subsurface information is taken into account when planning the well, close cooperation between the professional disciplines of drilling engineering, geology, and geophysics is necessary. For development wells (producer and injector wells), drilling and completion techniques should be applied to meet reservoir management demands and maximize

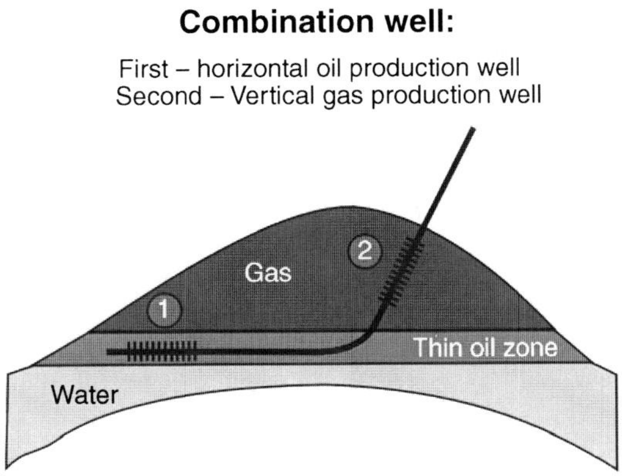

Figure 3.13: A sketch of a combined well design.

the recovery. Well design should be made with due consideration to the geology, technical issues, environmental issues, and license agreements.

3.3.1 Well design: geological conditions

The well prognosis will provide expectations for the following conditions as basis for designing the well:

- total depth;
- lithology;
- pressure;
- prediction of oil or gas.

3.3.1.1 Depth prognosis

The total depth prognosis for a well can be based on various criteria such as a requirement to reach a specific formation or trap, a particular seismic reflector, or a prescribed depth. Depth prognoses are made by converting available information from seismic time sections, and gravimetric and magnetic surveys to depth. This conversion will always contain some uncertainty which can be reduced by using nearby well information when available.

3.3.1.2 Lithology

Detailed prognoses of lithology anticipated to be present in the subsurface at the drill site are important in well engineering and drilling. Continued drill stem friction may impair the integrity of the wellbore and jeopardize logging results. Casing programs must be designed first not only for safety, but also to optimize the logging runs. This requires a detailed lithological prognosis based on all available information, including predictions of permeable and tight rock layers.

3.3.1.3 Pressure and depth

Well safety must never be compromised by the often-conflicting demands for technical results. To reduce the possibility of a "run-away" well, that is, losing control over pressure and flows, it is important to know where to expect overpressured or underpressured formations. Drilling muds are designed to contain formation pressure but can also inflict irreparable damage on reservoirs. Safe and efficient drilling is possible only with a good prediction of the expected pore pressure development with depth.

3.3.1.4 Anticipated fluid

Predictions of the fluids that may be contained in an undrilled prospect are generally uncertain. The techniques of basin modeling can help to reduce this uncertainty if information on source rock quality and temperature distribution in the basin is known. Modern seismics in many cases provide direct hydrocarbon indicators such as flat spots or bright spots if gas is present.

3.3.2 Well design: technical conditions

Technical inputs to well design include the following:

- purpose of the well: exploration, appraisal, injection, or production well;
- potential shallow gas or shallow water flow:
- reservoir objectives and targets;
- directional requirements;
- formation pore pressure gradient;
- fracturing pressure gradient;
- formation temperature gradient;
- expected lithological column;
- tectonic activity or any geological faults;
- expected fluids (gas, oil, water, CO_2, H_2S, etc.);
- strength and condition of top soils at seabed;
- any particular experience of local conditions;
- rig availability, equipment availability, and technical specifications.

3.3.3 Well design: environmental, regulatory, and contractual conditions

Sometimes referred to as "soft issues," requirements on well design arise from the following:

- regulatory requirements from authorities;
- company policies and procedures;
- fulfillment of license obligations;
- environmental issues: minimizing or avoiding impact in the particular area;
- considerations of promoting local business development.

3.4 Well completion

After the well has been drilled and casing has been set, the well must be completed. Completion of the production well is the process in which the well is enabled to produce oil or gas. Well completion should provide for the following in accordance with local conditions:

- safe and efficient production of hydrocarbons;
- provision of barriers to avoid leakage of hydrocarbons;
- allowing selective production of different reservoirs;

- provision of a means to shut down the well automatically;
- avoidance of sand production;
- avoidance of gas or water breakthrough;
- optimization of well performance;
- allowing intervention and repair work in the future;
- provision of interfaces for transferring hydrocarbons to production facilities.

3.4.1 The well

The well is created by drilling a hole (5–30 in. wide) into the earth with an oil rig turning a drill bit. After the hole is drilled, a metal pipe called a "casing" is cemented into the hole. In order to gain access to the reservoir, the casing and cement are either perforated (cased-hole completion) or an additional section of rock is drilled below the casing ("open-hole completion"). In most cases, several casings are set into the well, starting with a large shallow casing, and then deeper casings are set into smaller holes drilled through the upper casings.

3.4.1.1 Casing

Casing is installed in hydrocarbon wells to serve a multitude of functions. It is delivered in different sizes, weights, and steel grades to cater for different pressure regimes and chemical composition of formation fluids. The main functions of casing strings are:

- provide structural strength to the well in order to support all the weights and loads imposed during the lifetime of the well;
- seal off formations to ensure that drilling can continue to deeper targets;
- ensure that the well can withstand the anticipated pressures of any influx while it is being circulated out;
- seal off near-surface water aquifers to avoid environmental damage to the local water supply;
- act as a sealed conduit through which the production tubing can be installed to surface.

The size of the casing depends on the eventual tubing diameter required for optimum production characteristics of the well and ultimately the economics in the life cycle of the well.

3.4.1.2 Cementing

As the well is drilled, casing is installed and simultaneously cemented. Cementing is done to hold the casing in place. The drill pipe is removed from the hole, and heavy steel pipe (casing) is installed in the hole. This pipe is filled with cement and topped with a plug. Mud is then pumped into the casing while the cement is still wet. The pressure of the mud forces the plug and cement down inside the casing, so that the cement is forced into the space between the casing and wall of the borehole from the bottom of the casing (see Figures 3.3 and 3.14). The cement fills this space to the surface of the well. The task must be done quickly but carefully, as a poor cement job can adversely affect the finished wellbore. Once the cement has hardened, drilling may continue.

3.4.2 The completion process

The term "well completion" commonly refers to the process of finishing a well so that it is ready to produce oil or natural gas, or for an injection purpose as the case may be. The completion of a production well requires determining the characteristics of the intake portion of

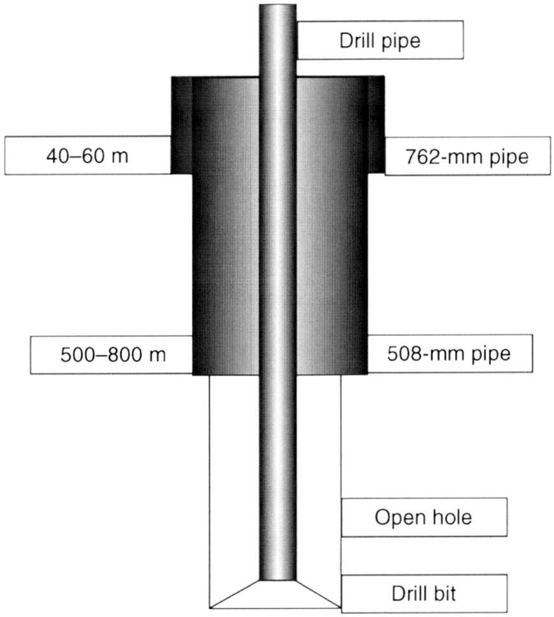

Figure 3.14: A sketch of an oil well, open-hole completion.

the well in the targeted hydrocarbon formation. It involves the final design of the well itself and the size of the production pipe to be installed.

3.4.2.1 Design criteria

Many factors influence the choice of completion method, including:

- purpose of the well: producer, injector, or disposal well;
- expected production potential of the well;
- type and quality of hydrocarbons;
- any corrosive constituents;
- reservoir pressure;
- future pressure depletion expectations;
- potential sand production;
- any needs to avoid specific reservoir zones containing water or gas;
- any future plans for recompleting or sidetracking the well;
- type of drilling technique: vertical, inclined, horizontal, or multilateral well;
- anticipated well re-entry and workover requirements.

Once the design basis is clear, the size and depth of the production tubing will be chosen. Depending on zonal isolation requirements, the installation of isolation packers between various zones may also be planned. Exact depths and requirements can often be determined only after the well has been drilled and logged through the reservoir. The productive zones can be identified to a high degree of accuracy from wireline logs, in order to ensure that the completion across the reservoir is optimized and targeted at the correct zones.

3.4.2.2 Types of well completion methods

Basically, well completions are either cased- or open-hole completions. Supplementing these two basic completion methods, it may be relevant to use sand or gravel screens, permanent completion, or multiple completions, depending on the characteristics of the hydrocarbon formations to be drilled. After finishing the well, production tubing is installed and the well is stimulated ready for production.

3.4.2.3 Open-hole completion

Open-hole completion is the most basic type, and is only used in compact formations which are unlikely to cave in. An open-hole completion consists of simply running the casing directly down into the formation, leaving the end of the piping open (see Figure 3.14), without any other protective filter. This type of completion is often used on formations that have been treated with hydraulic or acid fracturing. A sand screen or a gravel pack may be installed in the last drilled, uncased reservoir section. These tools serve to maintain structural integrity of the wellbore in the absence of casing, while still allowing flow from the reservoir into the wellbore. Screens also control the migration of formation sands into production tubulars and surface equipment, which otherwise can cause washouts and other problems, particularly in unconsolidated sand formations in offshore fields.

3.4.2.4 Cased-hole and conventional perforated completion

When the well has been drilled into the oil-bearing rock, the drill pipe is removed and the last string of casing is installed and cemented in place. A section of the casing that passes through the reservoir is then perforated—that is, holes will be created in the casing to allow oil and gas to pass into the casing, while the casing still provides a suitable amount of support and protection for the well hole. See Figure 3.15 for an illustration. The process of perforating the casing involves the use of specialized equipment designed to make tiny holes through

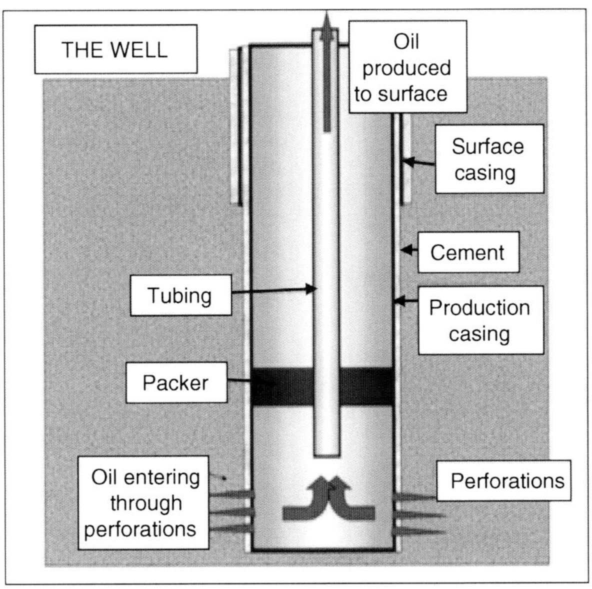

Figure 3.15: A sketch showing perforations in a cased-hole completion.

the casing as well as the cementing and any other barrier between the formation and the open well. In the past, "bullet perforators" were used, which were essentially small guns lowered into the well. The guns, when fired from the surface, sent off small bullets that penetrated the casing and cement. Today, "jet perforating" is preferred. This consists of small, electrically ignited charges, lowered into the well. When ignited, these charges poke tiny holes through the casing and cement to the formation.

3.4.2.5 Sand exclusion completion

Sand exclusion completions are designed for production in areas that contain large amounts of loose sand. These completions are designed to allow for the flow of natural gas and oil into the well, but at the same time prevent sand from entering the well. Sand inside the well hole can cause many complications, including erosion of casing and other equipment. The most common method of keeping sand out of the well hole is screening or filtering system. This includes analyzing the sand experienced in the formation and installing a screen or filter to keep sand particles out. This filter may either be a type of screen hung inside the casing or adding a layer of specially sized gravel outside the casing to filter out the sand. Both of these types of sand barriers can be used in open-hole and perforated completions.

3.4.2.6 Permanent completion

Permanent completions are those in which the completion and wellhead are assembled and installed only once. Installing the casing, cementing, perforating, and other completion work is done with small-diameter tools to ensure the permanent nature of the completion. Completing a well in this manner not only can lead to significant cost savings compared to other types, but also results in reduced flexibility for subsequent well recompletion.

3.4.2.7 Multiple zone completion, "smart well completion"

Multiple zone completion is the practice of completing a well such that hydrocarbons from two or more formations may be produced simultaneously, without mixing with each other. The technology was developed with the objective of placing inflow control valves in the completion string, so the production from the individual production zones in the well could be controlled, either in terms of full closure or alternatively adjusted up to the full flow potential of the intervals. It may be efficient in a horizontal well to add multiple completions to drain the formation most effectively. Achieving a complete separation is the main challenge here. In some instances, the different formations being drilled are close enough in nature to allow fluids to intermingle in the well hole. When it is necessary to separate different completions, hard rubber "packing" instruments are used to maintain separation.

This complex technology involves the installation of production sealing packers and remotely operated inflow control valves (electrically or hydraulically operated or a combination of both), and combining this hardware in a hot and high-pressure environment with the requirements of long lifetime and robustness. The required higher initial capital expenditures can often be justified by lower subsequent operation costs and a minimum of well interventions. Undesired water and gas breakthrough can be handled as the zones with high water production can be choked back or closed in without well intervention operations.

3.4.2.8 Installation of production tubing

After the casing has been perforated, tubing is installed. Tubing is used to bring oil and gas to the surface. It is a smaller diameter pipe that is lowered into the casing and held in place by packers which also isolate the producing layers of the rock (see Figure 3.15)—the production tubing; there are various nipples, valves, and accessories which provide safety barriers. They also simplify potential future workovers and re-entries into the well. In many cases, such accessories are used for production and pressure data gathering during the life of the well. The tubing hangs from a surface installation called the wellhead or *Christmas tree*. The wellhead includes valves, chokes, and pressure gauges, and makes it possible to regulate production from the well.

3.4.2.9 Well stimulation and hydrocarbon lifting

After a flow path is established, acids or other fluids are often pumped into the well to clean (enlarge old channels), fracture (to create new ones), or otherwise prepare and stimulate the reservoir rock to optimally produce hydrocarbons into the wellbore. The smaller diameter of the tubing produces hydrocarbons at an increased speed, in order to overcome the hydrostatic effects of heavy fluids such as water. In most wells, the natural pressure of the subsurface reservoir is high enough to push the oil or gas all the way to surface. However, this is not always the case, especially in depleted fields where pressures have been lowered by other producing wells. Common solutions are submersible (downhole) pumps, gas lift, or surface pump-jacks. The use of artificial lift technology is often referred to as "secondary recovery."

3.5 Well testing

After the well is drilled and logged, its producing capability should be tested. At this stage, there may be a temptation to cut costs by not testing properly. Such cost savings can prove to be costly in the end.

3.5.1 Data acquisition during the drilling process

Recent developments in the measurements while drilling (MWD) technology have given the possibility to detect and communicate formation characteristics to surface during the drilling process. The data may be gamma ray, resistivity, neutron porosity, and sonic information. This is commonly described as logging while drilling (LWD). Surface access to these real-time downhole measurements is an important aid for the drilling process and further planning of the drilling operations as these petrophysical data are used as input parameters for further steering of "where to drill the well," often referred to as "geo-steering," and in the decision-making process of when to stop further drilling of the well.

Well test analysis is an advanced and well-documented methodology that in competent hands yields important information about the reservoir's potential, its boundaries, fault structures, etc.

3.5.2 Wireline logs

Wireline logging remains the main tool for evaluating subsurface formations. After each section of the hole is drilled and before casing is run into the hole, the wellbore is logged. Wireline logs can be separated into four functional groups, which measure the following:

- resistivity;
- porosity/lithology;
- production characteristics;
- miscellaneous parameters.

Wireline logs measure correlation, bed thickness, porosity, and water saturation. This allows conclusions to be drawn on lithology, permeability, the presence and type of hydrocarbons, and the mobility of hydrocarbons.

3.5.3 Core sampling

Coring operations bring large rock samples to the surface. Because the operation is time consuming and costly, conventional cores are normally only taken in potential reservoirs in order to investigate the geological model and measuring reservoir properties such as:

- rock parameters: porosity, permeability, residual fluids, and lithology;
- prediction of gas, oil, or water production;
- aerial changes in porosity, permeability, and lithology required to characterize the reservoir for reserve estimates and reservoir modeling;
- irreducible water saturation;
- unaltered wettability and/or saturation state for special core tests;
- directional permeability studies;
- calibration of, and/or improved interpretation from, downhole logs.

3.5.4 Testing the fluid flow

In well testing, the fluids in the formation are allowed to flow to the surface. The objective of the test is to determine the productivity of the well and to collect samples during the flow periods. Downhole pressure will be recorded during the test period. Subsequent analyses of the pressure data will yield information about reservoir pressure, permeability and productivity, and formation damage.

Another objective of well testing is to obtain samples of produced fluids, oil, gas, and water. These samples are collected by either bottom-hole sampling or surface sampling techniques. Depending on the reservoir characteristics, it may be recommended to sample the fluids on the separator. The most crucial factor here is to obtain a representative and correct gas to oil ratio (GOR), and to ensure that the well flow is stable while sampling. Samples for pressure, volume, and temperature (PVT) analyses are taken from the gas and oil stream. These PVT analyses should be carried out according to well-established standards and procedures. The PVT report provides important information for the subsequent reservoir evaluation process.

3.5.5 Testing samples for fluorescence

Samples of rock containing hydrocarbons can be examined under ultraviolet light for evidence of fluorescence. Ultraviolet light (wavelength 3600 Å = 3.6×10^{-7} m) causes fluorescence of almost all crude oils over the entire range of gravities. The color of fluorescence provides an indication of the gravity of the crude oil.

3.5.6 Tasks of the well site geologist

Drilling operations are often attended by a geologist, who will be responsible for the proper testing of hydrocarbon shows and for monitoring whether the formation characteristics conform to expectations. Geologists will be involved in the following tasks:

- analysis of the cuttings;
- gas determination from the drilling mud;

- hydrocarbon evaluation;
- wireline logs;
- core sampling;
- drill stem testing.

3.6 Drilling organizations

Drilling organizations have tended to be strongly hierarchical and regarded as having an old-fashioned style. Such organizations include multiple layers of management, with the operator of the field or license being ultimately responsible for all decisions related to the operations. The main reason for this is the operator's responsibility for and need to control safety, and commercial and reputational risks. Most providers of drilling rigs and services have worked on a day rate basis, with limited knowledge of the total well construction process.

Lately, many operators have welcomed more comprehensive collaboration with service providers and rig personnel's input in the planning and decision process. This has many times given excellent results. This involves pushing decision-making authority down into the organization, allowing the experts from particular disciplines better insight in the overall process and giving them the opportunity to do a better job. The end result is a flatter organizational structure with more intent on giving room to maneuver for the various players, as well as trust and confidence in the work provided. The general view in the industry is that such "flat" organizations tend to be more effective than the layered models employed previously. See Figure 3.16 for a general outline of a drilling organization.

A problem in the organizations has often been in the interface between drilling/completion personnel and their customer, which is the reservoir or production engineering group of a particular project. One reason for this is that the drilling and completion group carries out operations to implement plans that have been made elsewhere in a company. The group then takes ownership of budgets and targets of a well, and is responsible to carry out the operations within budget and on time. This sometimes leads to differing objectives between customer

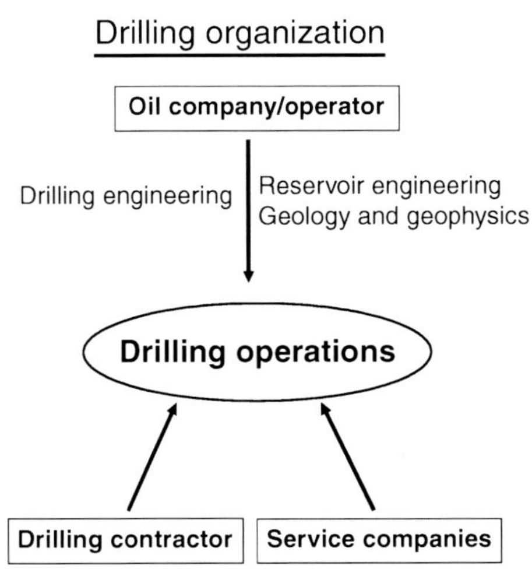

Figure 3.16: Drilling organization outline.

(reservoir/production engineering) and the executing drilling organization. Some operators or projects now organize their drilling and completion work such that all disciplines are represented in one team, breaking down the traditional barriers.

3.6.1 Contracting a drilling service provider

Drilling services and rigs are contracted in three ways: as a day rate contract, an incentive contract, or a lump sum contract. Different situations, legislation, operational challenges, and project risks provide the basis for selecting the type of contract that should be employed. Often, however, the contract type selected is based on the experience from key operator personnel and/or to suit a company-wide contracting strategy. A key responsibility of the operator is to have a contracting strategy which is suitable for the operation at hand, and is clear to the parties involved.

A total suite of contracts for a drilling and completion operation tends to employ elements of all three types of contracts mentioned above. However, one particular type is generally selected as the main contracting strategy. The successful application of a contract is that the personnel involved have an active approach to its use, and know its content. Furthermore, a successful contractual relationship requires mutual trust and openness between the contract parties.

3.6.2 Drilling performance monitoring

The performance of drilling and completion personnel needs to be monitored against the following criteria:

- HSE performance (i.e., the avoidance of injuries and health hazards, accidents, and pollution);
- meters per day progress;
- costs per meter and efficiency (uptime/downtime) of the rig operation.

The above are the factors through which individual and team performance is measured. For each well and operation, an efficient organization should have progress criteria set with targets measured against similar wells or operations.

3.7 Health, safety, and environmental requirements

Drilling rigs are inherently connected with safety-related risks. Heavy machinery is in use, personnel are required to handle large pieces of equipment, and the work environment is often considered to be rough and dirty. Increased focus on the above issues has contributed to safer work environments in drilling. Safety management systems are today an integral part of any drilling operation, and the HSE standards have been continuously improved through training and focus on HSE issues. In many areas, these standards are strictly regulated through legislation. In other areas, the legislation may be poor. Most operators have developed their own internal requirements to safety standards of equipment and systems, and they are therefore implemented on a worldwide basis.

Cuttings and chemicals used in drilling fluids and their discharge are the main environmental concerns in the drilling operation. Operators are focusing on reducing the amount of emissions or reducing the toxicity of the chemicals used. This has led to introduction of on-site cuttings-cleaning equipment, downhole disposal of cuttings, central cuttings-destruction plants, cementing of cuttings in order to make them inert, introduction of new environmentally friendlier muds, and similar.

A particular environmental concern has been with pollution originating from OBM. This has resulted in the introduction of more environmentally benign drilling muds as reviewed in an earlier section.

3.8 Summary

Drilling boreholes is a direct method of investigating subsurface geology although it is expensive and provides information only at discrete locations. Drilling also creates access to one or several reservoirs at some depth below the surface, and hence provides a flow path of hydrocarbons for production in the exploitation phase of a field development. Careful well design and well completion will result in good reservoir coverage and exploitation. It will also allow a reduced number of drilled wells and consequently reduced cost and higher profit. HSE must also be taken into account during planning and operation of drilling activities in order to achieve zero harm to people and environment.

Flow assurance

This chapter addresses the transfer of hydrocarbons from a reservoir to the points where they can be processed. Offshore production environments present particular challenges for moving hydrocarbons, not least because the high cost of processing facilities offshore provide a motive for transporting untreated wellstreams across substantial distances to centralized (and ideally onshore) processing facilities. Many of these challenges have been addressed in Norwegian offshore oil and gas production, and the chapter therefore draws extensively on examples from this region.

4.1 The need for flow assurance

Flow assurance refers to the safe and uninterrupted transport of hydrocarbons from a reservoir to processing facilities. The hydrocarbons flowing through a producer well are often a mixture of oil and gas accompanied by water, various gases, and solids, all of which must be transferred to a point where they can be adequately separated and handled. Flow assurance often involves *multiphase transport*, which is the simultaneous flow of gas, oil, and sometimes

Figure 4.1: The Snøhvit subsea to land development. Drawing: Even Edland, Statoil.

solids through a pipeline. It covers the entire spectrum of design tools, methods, equipment, knowledge, and professional skills needed to transport different reservoir fluid phases simultaneously in pipelines.

The basis for effective multiphase transport and design lies in high-quality well fluid sampling and characterization, advanced fluid mechanics, flow simulation tools, and in-depth understanding of fluid behavior. Numerous complications arise from liquid slugging, sand production, and the simultaneous occurrence of physical–chemical flow-impeding phenomena such as wax, gas hydrates, scale, asphaltenes, and emulsions. The reservoir gases carbon dioxide (CO_2) and hydrogen sulfide (H_2S) also create noxious chemical mixtures, which can seriously corrode hardware. All these conditions of a particular field have to be well understood for the purpose of preventing and treating the flow problems which can potentially arise.

Multiphase transport solutions have been implemented for the following fields (examples from the Norwegian offshore):

- Huldra to Heimdal fields (platform to platform);
- Midgard to Heidrun fields (subsea to platform);
- Troll A platform to onshore processing facility of partly processed fluid;
- Snøhvit field (subsea to onshore processing facility of full wellstream) (Figure 4.1).

4.2 Pipeline transportation modes

Pipeline transportation of gas and liquids can be configured in different flow modes:

- as single-phase gas or liquid flow;
- as multiphase (including two-phase) gas and liquid flow;
- as gas and liquids in dense phase (supercritical flow).

Flow assurance issues relevant to these pipeline transportation options are briefly discussed under the following subsections.

4.2.1 Single-phase gas transportation

Single-phase gas transportation is the movement of gaseous hydrocarbons in a pipeline, where the occurrence of liquids should be avoided. Hydrocarbon and water dew points are critical factors when pipelining single-phase gas, that is, transmission and distribution of typical sales gas.

In large gas pipeline systems which handle gas from a multitude of sources, certain specifications are defined in order to avoid precipitation of liquids which would cause problems in the pipelines. The *dew point* of a gas at a certain pressure is the temperature at which liquids will start to form. For North Sea gas, the following dew point specifications are common:

- hydrocarbon dew point at 1–69 bar pressure: 3 °C;
- water dew point at 69 bar pressure: 12 °C.

The consequences of not meeting the dew point requirements may be severe, due to slugging in the case of hydrocarbon liquids fall out and risk of hydrate build-up inside a pipeline system in the case of excess water.

4.2.2 Single-phase liquids transportation

Single-phase liquids transportation is the movement of crude oil or another hydrocarbon liquid in a pipeline, where the occurrence of gas should be avoided.

When transporting crude oil or condensate in a pipeline, it is often possible to include large amounts of gas, particularly heavier gas components (NGLs) with the liquid flow. Such crude oil or condensate can still be transported in a single-phase pipeline provided that sufficient pressure is maintained in order to avoid the formation of free gas. This requires that pressure throughout the pipeline is maintained above the vapor pressure of the liquid.

Another issue with liquids transportation is the control and removal of wax that tends to precipitate in a pipeline carrying crude oil (Section 4.3.5).

4.2.3 Two-phase flow

In some cases, it may be relevant to undertake an initial processing of a wellstream on a platform or other processing facility in order to remove water and solid particles, and then transport the resulting hydrocarbon liquids and gas together in a pipeline system.

When gas and liquids (typically condensate) flow concurrently in separate phases in a pipeline, liquid build-up in low spots along the pipeline route may cause flow restrictions. To avoid or minimize the effects of this, special sweeping pigs are regularly run through the pipeline, to push out accumulations of liquid.

4.2.4 Multiphase oil systems

A multiphase oil system is the simultaneous transport of liquid hydrocarbons, gas, water, and solid particles in a pipeline, where liquid hydrocarbons are the predominant flow component. Multiphase oil systems are found at oilfields connecting wellheads with a central processing unit either onshore or offshore.

The main challenge for oil systems is the accurate prediction of pressure drops along the pipelines, properly taking fluid properties into account (e.g., the oil–water flow regime, emulsification, and viscosity). These aspects are vital for appropriate selection of pipeline diameters, pipeline insulation, and heating, and ensuring that separators receiving the flow are properly dimensioned to receive liquid surges (slugs).

4.2.5 Multiphase gas/condensate systems

Multiphase gas/condensate systems are pipeline transport systems carrying hydrocarbon liquids and gas as well as water and solids, but with gas being the predominant flow component.

In gas/condensate systems, it is important to predict the volumes of condensate and water accumulating in the pipeline. Also, the severity of liquid surges must be determined. Such predictions are especially important for designing *slug catchers*, which are high-pressure, liquid receiving and storage facilities (Figure 4.2), and for evaluating the impact of liquid surges on downstream processing equipment.

Figure 4.2: The slug catcher at Melkøya for the Snøhvit onshore facilities, Norway. Photo: Statoil.

4.2.6 Dense-phase transportation of gas and liquids

Figure 4.3 shows the pressure–temperature relationship for a rich gas blend from Statfjord and adjacent oilfields. This blended gas stream originates as associated gas from oilfields, after separating the wellstream into stabilized crude oil (suited to transport by tanker ships) and a gas stream containing substantial amounts of natural gas liquids.

This phase curve represents the bubble point line and the dew point line of the mixture. Inside the curve (phase envelope), a two-phase equilibrium between gas and liquids exists, while at any temperature outside the envelope the mixture is in single phase, either as liquid, gas, or dense phase. Dense phase is a state resembling a vapor and occurring at pressures above the cricondenbar, which is the maximum pressure at which two phases can exist.

This particular dense-phase behavior of a rich gas mixture is the basis for bringing rich gas from an offshore field to an onshore processing plant, or likewise onshore, from a remotely located field to a central processing facility, to avoid flaring of associated gas.

The gas stream represented in Figure 4.3 is brought to shore in a 30-in. pipeline. Pressure and temperature of the gas drop along the pipeline as indicated by the red line. The most crucial flow assurance issue when operating a dense-phase pipeline is to stay above the cricondenbar for the actual rich gas mixture. Otherwise, condensation of liquids and slugging may upset the gas flow.

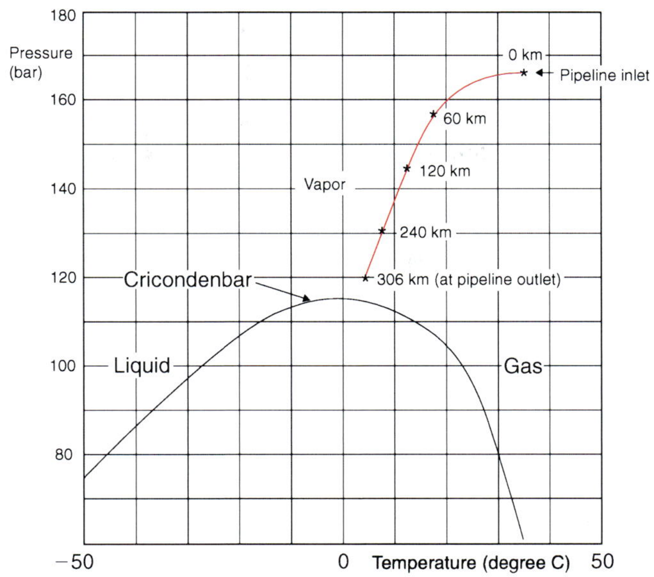

Figure 4.3: Dense-phase pipeline flow. Example: 30-in. rich gas pipeline from Statfjord (NCS) to Kårstø (Norway). *Source*: Drafted by K. Traa for the present book.

4.3 Common issues in flow assurance

Slugging, sand production, fluid phase behavior, asphaltenes, wax, emulsions, gas hydrates, scale, and corrosion are issues that can be potentially disruptive to oil and gas production unless properly dealt with.

4.3.1 Slugging

Slugging refers to irregular flow and surges of gas and liquid through any section of a pipeline. It is a flow-impeding phenomenon. Slugs may be generated in several ways:

Hydrodynamic slugging: Hydrodynamic slugging may develop when liquid waves grow on the gas–liquid interface and eventually close the cross-section, thus forming liquid slugs.

Riser slugging: Riser slugging occurs when liquid blocks the low point where a down-sloping pipeline is attached to a riser. The blockage initiates the slug, which then grows upward in the riser and back through the pipeline. This continues until the pressure build-up over the slug is sufficiently high to blow it out of the riser.

Terrain slugging: Terrain slugging involves slug development where a pipeline traverses through rough terrain (seafloor or onshore). The slug picks up liquid accumulated in inclined sections and may become very extensive.

Transient slugging: Transient slugging is liquid build-up in pipelines due to increased liquid flow rates at pipeline exits to processing facilities in response to changes in operating conditions.

The most common way to handle heavy slugging is by installing slug catchers at onshore processing plants, or making allowances for extra liquid storage volumes in the inlet separators on offshore processing platforms. These facilities are designed to receive considerable quantities of liquid over short periods of time. An alternative is to use static choking where risers enter processing facilities. Topside choking helps eliminate riser slugging by increasing the pressure drop across slugs. The slug is forced back down the riser until a more stable, bubbly gas/liquid flow is established.

Each of these methods has disadvantages, as slug catchers are expensive and space demanding and static choking reduces the production rate. One response to this has been the development of active slug control. Here, continuous flow and pressure measurements are used to automatically adjust the opening of a topside choke to keep slugging in check. By introducing faster control systems, it is possible to attain a relatively constant flow through the topside choke.

4.3.2 Sand production

Sand production refers to the dislodging of small rock particles and individual sand grains from reservoir rocks during hydrocarbon production, and their subsequent incorporation in the wellstream. With the steady increase in use of subsea production systems, where it is difficult to intervene if problems arise, it is essential that the sand behavior can be predicted. The key parameters for this prediction are:

- the threshold velocity, below which there is no sand transport and thus there is a danger of the sand accumulating as a stationary dune;
- the considerably higher flow velocity necessary for all of the sand produced to be transported in suspension.

Although the simultaneous flow of fluids and solid materials is common in many industrial applications, modeling is still a relatively immature science. Sand production is especially prevalent in production from oil-dominated reservoirs.

4.3.3 Phase behavior of petroleum fluids

To be of value, the simulation of multiphase flow and certain physical–chemical phenomena must be based on representative fluid samples and correct fluid characterization. Although today's simulation models are very reliable, there are still areas requiring improvement and validation, for example, asphaltene modeling and hydrocarbon–aqueous phase equilibrium in the presence of brine and hydrate inhibitors.

Figure 4.4 shows an oil phase diagram (note that this figure is identical to Figure 1.3). Depending on the design and operation of the production and transport system, some or all of the phase boundaries shown in the diagram may be crossed as oil is produced from the reservoir. As the oil flows from the reservoir toward the wells, its pressure and temperature decrease. As the pressure and temperature reduce, asphaltenes start to separate from the liquid oil solution; we could have drawn an *asphaltene precipitation envelope* (APE). As the pressure and temperature are further decreased, the *bubble point line* (thick curved line) is reached, below which gas starts to evolve from the solution. Hence, from this point on, a two-phase flow regime is present. Each of the thinner curved lines below the 100% liquid line represents a 10% increase in evolved solution gas, by volume. It is seen from the diagram that if the reservoir temperature is sufficiently high, all hydrocarbons may be in the gas phase at initial reservoir conditions. This is the case in many gas or gas-condensate reservoirs.

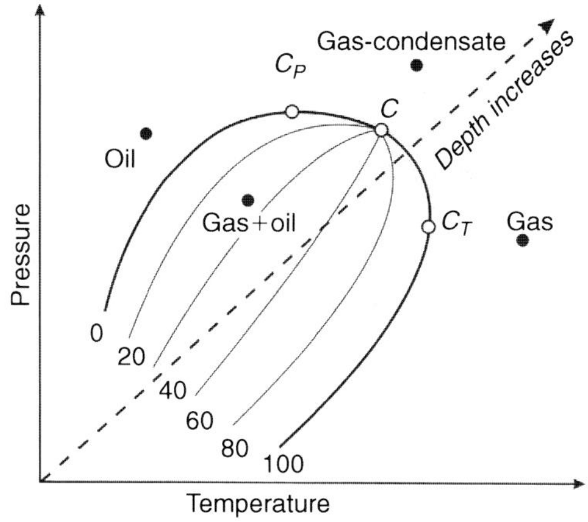

Figure 4.4: Oil phase diagram.

As the fluids reach the wellhead, the pressure may be so low that gas lift or a booster pump may have to be used in order to make the well flow. From the wellhead, the hydrocarbon fluids flow through pipelines to the inlet separator at the processing facilities, during which the pressure and temperature are lowered further. The wellstream may then cross the *wax appearance temperature* (WAT) (not shown in the diagram). As this line is crossed, wax begins to form. The wellstream may also cross the *hydrate formation line*, below which natural gas and water will combine to form gas hydrates. Asphaltenes, wax, and gas hydrates are further explained in the following sections.

It is worth to note that the occurrences of flow-impeding physical–chemical phenomena such as asphaltenes, wax, and gas hydrates are highly dependent on the pressure and the temperature of the wellstream. Hence, the fluid samples taken from an exploration or appraisal well must be compared to the actual pressure–temperature regime along the flow line to determine whether these phenomena will represent problems for field development.

4.3.4 Asphaltenes

Asphaltenes are highly polar, polyaromatic hydrocarbons, which are the heaviest components in crude oils and give rise to the oils' characteristic black color. Asphaltenes may aggregate (flocculate) and precipitate in response to changes in chemical composition, pressure, or temperature. Both precipitation and deposition may occur in the matrix of reservoir rocks, in wellbores, and in transport systems. This may lead to reduced productivity, costly removal procedures (workovers), or, at worst, a complete halt in production. Naturally, it is important to have knowledge of asphaltene chemistry and behavior to predict where problems may occur and to find ways of preventing and handling them.

The most common method for asphaltene control is to use solvents, especially in wellbore tubing, whereas the most common method for prevention is to add asphaltene inhibitors, usually dispersants restricting aggregate growth. Ideally, an inhibitor should be effective, modestly priced, and environmentally acceptable.

4.3.5 Wax

Wax is a class of hydrocarbons and a natural constituent in most petroleum fluids. Typical problems arising from wax precipitation include deposition on cold surfaces, such as pipeline walls, increasing oil viscosity until the oil may ultimately gel, as well as sedimentation in storage tanks.

Most crude oils tend to deposit wax on the inner pipe wall, which must be scraped off at regular intervals, to avoid build-up of flow restrictions. Wax removal is done by running special scraper pigs through the pipeline, pushed by the flow (Figure 4.5). Such pigs utilize a combination of scraper discs and an internal bypass system. For this to be possible, the pipeline must be equipped with pigging facilities at both ends.

Wax deposition is a major obstacle for long-distance tie-backs from satellite oilfields. Present solutions involving insulation (to keep operating temperatures above the WAT) and pigging may not be viable or may prove to be too expensive for subsea production systems. Furthermore, chemical inhibition is still an inadequate alternative. However, research has shown that a controlled cooling or *cold flow* concept has considerable potential. This concept concerns the deliberate promotion of wax precipitation by rapid cooling upstream of a multiphase pipeline. The result is a suspension of wax crystals which can be easily transported when the oil is at the same temperature as its surroundings throughout the entire length of an uninsulated pipeline.

4.3.6 Emulsions

When water and oil flow together in a pipeline, one of the phases tends to be dispersed as droplets in the other phase. This dispersion is often called an emulsion, and in oil production water is most often the dispersed phase (i.e., water droplets occur in the oil). Furthermore, emulsions are stabilized by naturally occurring components in oil, such as asphaltenes, resins, and organic acids, and even solid particles like wax, scale, and fine mineral particles may act as stabilizing agents. This, of course, also complicates the separation of produced water from oil, a procedure which has to meet stringent specifications concerning the maximum water content allowed in export oil, and the maximum amount of oil and other contaminants allowed in water if it is to be discharged into the sea. Also, the viscosity of water-in-oil emulsions can become significantly higher than that of the oil itself, resulting in an increased pressure drop along a pipeline and reduced transport capacity.

Figure 4.5: Example of scraper pigs. *Source*: Pigtek Ltd.

Remedial treatment, that is, the improved separation of emulsions, can be achieved in three ways:

- chemically by adding demulsifiers;
- mechanically in large separation tanks, usually at high temperatures;
- electrically (electro-coalescence).

Sufficient knowledge now exists to show that the addition of a specific demulsifier at subsea wellheads will substantially reduce the effective viscosity experienced in multiphase flow lines. Recent research has enabled optimal choices of chemical demulsifiers, while ongoing work is concerned with the fundamentals of electro-coalescence and the development of a compact electrostatic coalescer.

4.3.7 Gas hydrates

Resembling snowballs or slush, gas hydrates are crystalline substances comprising a framework of water molecules whose cavities are occupied by a particular gas or mixture of gases. Gases may include methane, ethane, butane, and inorganic molecules such as CO_2 and H_2S. Gas hydrates may be generated when hydrocarbon molecules come into contact with water, typically at temperatures below 15–25 °C and at elevated pressures.

The forms which gas hydrates take during pipeline transport can vary from finely dispersed particles, which are easily carried along with the flow, to large cohesive lumps capable of causing partial or complete blockage. The latter can be a serious impediment to production, as it may take weeks or months to clear them. Without proper design and adequate operating procedures, hydrates would form naturally in most North Sea multiphase systems, such as pipelines, risers, wells, and processing equipment, due to the cooling effect of the surroundings. Water temperatures in certain deepwater and arctic regions can approach −1.8 °C, which is the freezing point for salt water, augmenting the risk of hydrate formation. Many efforts are therefore devoted to predicting their occurrence, preventing their development, and safely removing them.

The usual way of controlling hydrates is by adding thermodynamic chemical inhibitors (i.e., methanol and glycol). Considerable efforts have been made over the years to optimize their use. Essential to this work is the extensive testing and analysis of the basic properties of monoethylene glycol (MEG) and inhibitor phase distributions. In addition, operational investigations are carried out to minimize the need for applying chemicals.

An alternative to using chemicals is to apply heat. Extensive studies concerning the development and qualification of various heating methods have led to *direct electric heating* (DEH) of flow lines being adopted at the Åsgard and Huldra fields in the Norwegian Sea and the North Sea, respectively. Pipe bundles heated by warm water are used at the Gullfaks satellites to reduce the use of chemicals and utilize waste heat. Bundles comprise flow lines, hot water pipes, and control signal umbilicals encased in a large-diameter external pipe.

4.3.8 Scale

Oilfield scale mainly comprises inorganic salts, the most common being calcium carbonates and barium and strontium sulfates. They may form from brine (formation water) as it undergoes changes in pressure and temperature, or where two incompatible fluids are inter-mingled (e.g., formation water rich in calcium, strontium, and barium mixed with seawater rich in sulfate). The former mechanism generally gives rise to carbonate scales while the latter generally produces sulfate scales. Scale problems can arise in various situations:

- during drilling and well completion, if the drilling mud or completion fluid is incompatible with the formation water;
- at the commissioning stage of new injectors, if the injection water is incompatible with the formation water;
- during production, when a well starts to produce formation water with the hydrocarbons;
- during wellstream processing, when significant quantities of produced water put process equipment at risk.

Commingled production (i.e., where wellstreams from various formations, reservoirs, or individual wells are mixed together) can make the problem worse. Scale can be removed using dissolvers. However, sulfates are particularly resistant and may only be cleaned from surfaces using special chemicals (chelants) and/or by mechanical means. Carbonate scale is usually the easiest to handle as carbonates are readily dissolved by hydrochloric acid, acetic acid, and the like.

4.3.9 Corrosion

Because of high CO_2 contents and elevated temperatures, untreated or only partly treated well fluids containing water can corrode carbon steel.

Ways of overcoming the corrosion problems are:

- addition of film-forming corrosion inhibitors that will protect carbon steel;
- use of corrosion-resistant steel alloys.

Compared with earlier prediction and inhibition practices, recent developments in modeling and increased inhibitor efficiency have made it possible to extend the application domain for carbon steel, which is the most widely used pipeline material today. This can be an important contribution to cost-effective, competitive field development solutions. The final selection of materials should be based on a life cycle cost analysis of various alternatives.

4.4 Challenges in flow assurance

There are still a number of major challenges in flow assurance, on which further research efforts are expected. They include long-distance oilfield tie-backs, subsea processing and flow assurance, deepwater flow assurance, and safety and environmental protection.

4.4.1 Long-distance oilfield tie-backs

Long-distance multiphase transport is becoming increasingly important for regional developments involving extensive tie-ins of satellite fields to existing production centers (e.g., the Åsgard and Norne areas). So far, however, ultra-long multiphase pipelines have only been installed for gas/condensate systems. Nevertheless, there are environmental and economic incentives to develop oilfields in a similar fashion as the Snøhvit gas/condensate field, that is, by using subsea installations only. In this particular development, the wellstream flows 143 km from the Snøhvit subsea facilities to the LNG facilities at Melkøya near Hammerfest.

It will also be desirable to significantly extend the reach for oil-dominated systems from the present 70 km. Flow assurance challenges for this purpose will mainly concern fluid control as wellstreams cool down, thereby increasing the risk of precipitation and pipeline plugging by hydrates, wax, etc.

Another aspect concerns the prediction and remediation of slugging where pipelines and risers enter processing facilities. This is particularly important for risers linking deepwater subsea production systems to ships or platforms. The impact of produced sand on multiphase transport will also have to be investigated.

4.4.2 Subsea processing and flow assurance

The use of seabed separators to remove produced water from wellstreams is expected to increase in the near future. This approach has many advantages, not least for facilitating regional developments, and re-injecting produced water and releasing more topside space on host production vessels to process larger volumes of oil and gas. It will also help to mitigate certain important flow assurance phenomena. There is a need to better understand how fluids behave in such equipment, and how this equipment interacts with wells, flow lines, and receiving facilities.

4.4.3 Deepwater flow assurance

Irrespective of whether fields are developed in shallow or deep water, the flow assurance challenges are broadly the same, even if pressure and temperature regimes are somewhat different. The objective is to avoid faulty designs or procedures that may lead to severe problems as deepwater intervention is notoriously difficult and complex. At the same time, it is vital that the cost per barrel of oil is kept at an acceptable level. Success lies in ensuring that deepwater flow assurance solutions yield high regularity, minimal downtime due to precipitation or plugging, and safe and efficient remediation.

Deepwater developments pose special challenges in certain important areas such as slug and hydrate control, the latter being due to the lack of effective depressurization. At present, the most widespread method for slug control is to inject gas at the foot of a riser. In the future, novel techniques are likely to emerge for hydrate remediation and slug control.

4.4.4 Safety and environmental protection

Many aspects of flow assurance remediation have long been based on chemicals. Reducing the number of chemicals in use and reducing their environmental impact as much as possible will be necessary in order to meet increasingly stringent demands for environmental protection.

Ways must also be found to effectively handle the increasing volumes of produced water from maturing giant oilfields to meet the aim of "zero harmful discharge."

In offshore environments, fisheries and petroleum activities must coexist constructively so that frontier areas of today become the prosperous hydrocarbon provinces of tomorrow.

4.5 Summary

Flow assurance refers to the safe, uninterrupted, and simultaneous transport of gas, oil, and water from a reservoir to processing facilities. Flow assurance usually involves multiphase transport, requiring relevant design tools, methods and equipment, as well as knowledge and professional skills.

Flow assurance aims to solve problems concerned with flow-impeding phenomena such as slugging, sand production, asphaltene precipitation, wax precipitation, emulsification and formation of gas hydrates, scaling, and corrosion. The main challenges in flow assurance are associated with long-distance oilfield tie-backs, subsea processing, deepwater aspects, and environmental aspects.

Processing requirements in oil and gas production

The wellstream from a hydrocarbon reservoir is rarely of a quality that is suitable for commercial trade. Rather, it is a mixture of oil, gas, and water as well as several contaminants. An oil and gas development project must incorporate processing of the wellstream to obtain products that can be sold and transported.

The re ning industry and the petrochemical industry are two major components of the overall petroleum industry, which process oil and gas toward products that consumers can use. These downstream industries are beyond the scope of this book, but some aspects of the interface between upstream oil and gas production and the downstream activities will be brie y mentioned in this chapter and the next.

5.1 The products from oil and gas production

Following is a brief overview of terminology applied to crude oil, natural gas, and by-products at various stages of processing:

Crude oil: Crude oil is extracted from reservoirs as a liquid, and consists of a wide range of hydrocarbon molecules ranging from gases (C4 and lower) to large, heavy molecules. Commercially important crude oil qualities include the gravity (measured in API degrees) and the sulfur content. Crude oil is converted to petroleum products in re neries.

Petroleum products: This term is usually applied to products manufactured in oil re neries, using mainly crude oil as re ning feedstock. Gasoline (petrol), gasoil (diesel and light heating oil), kerosene, jet fuel, naphtha, and heavy fuel oil are important petroleum products.

Condensate: Condensate is a stable hydrocarbon liquid which is obtained from processing gas, in which the condensate has been dissolved. Condensates are generally light liquids with a water-clear or yellowish appearance.

Dry gas (sales gas): Natural gas that has been processed to meet sales speci cations, that is, the product known to consumers of natural gas. It contains mainly CH_4 (methane), C_2H_6 (ethane), and limited quantities of heavier gases. Water and carbon dioxide (CO_2) content is limited.

Rich gas (field gas): Natural gas as produced at eld installations prior to gas processing. It contains substantial amounts of natural gas liquids (NGL; C2 and heavier, the wet part) in addition to the methane. The water content must be within limits required by the transportation system.

NGL: Ethane and heavier hydrocarbons which are produced as a part of the gas at eld installations, but are separated out in gas processing and usually sold in liquid form. The sales products are ethane (C2), propane (C3), butane (C4), and condensate (C5+).[1]

[1] Hydrocarbons are often referred to by the number of carbon atoms in their molecules, that is, C2, C3, C4, etc. Their chemical formulae are: ethane, C_2H_6; propane, C_3H_8; butane, C_4H_{10}.

Liquefied natural gas (LNG): Natural gas (dry gas) that has been lique ed by cooling for the purpose of storage and transportation. The storage is at atmospheric pressure and a temperature of about −162 C. LNG consists mainly of methane (C1), with small amounts of ethane (C2). Other gas components must be removed in the process of making LNG, as they would cause problems at the low temperatures.

Liquid petroleum gas (LPG): Propane (C3) and butane (C4). These are relatively heavy gases which are lique ed for the purpose of storage and transportation. As liquids they are stored either under pressure at ambient temperature or at atmospheric pressure and low temperatures.

5.2 Processing requirements

The type and extent of processing required for the wellstream depend on the speci cations (required qualities) for the sales product, and the properties of the wellstream obtained from the reservoirs. Other important conditions are the requirements concerning injection into reservoirs, transportation and storage of product, environmental protection, and operations.

5.2.1 Sales product specifications

Sales contracting and required product qualities for oil and gas are reviewed in Chapter 6. They are an important determinant of the need for processing oil and gas prior to delivery to customers. Oil and gas producers need to deliver products to buyers at agreed qualities and rates. Processing facilities need to be planned and implemented so as to permit the negotiation of sales contracts which preferably can be in conformity with established industry practices, and which the company will subsequently be able to honor.

Sales speci cations for oil and gas are generally designed so as to also ensure the product s suitability for transport. For oil transport by tanker ships, crude oil must be *stabilized*, that is, have a vapor pressure usually not exceeding 0.8 bar, which requires removal of most gas components from the crude oil. Furthermore, the crude oil must be processed at the producing installation so as to ensure low levels of bulk impurities and certain corrosive components such as hydrogen sul de (H_2S). Most re neries take the crude oil through a process for removing impurities (e.g., a desalter) prior to entering the sensitive re ning processes. The chosen mode of transport (pipeline or ships) has some implications for what can be acceptable quality, particularly in terms of vapor pressure, as will be reviewed in the next chapter.

Pipeline gas sales contracts include requirements on the producing eld to supply gas at prescribed rates, and severe economic penalties may apply if the producer fails to meet the contract terms. This is especially true for pipeline gas contracts, which usually specify a *daily contracted quantity* (DCQ) and maximum and minimum rates in the buyer s option. Gas supply availability is also speci ed in the gas contracts (Jahn et al., 1998).

Following is an example of standards that may apply to natural gas sold by pipeline in Europe:

- gross calori c value within 38 40 MJ/Sm^3;
- CO_2 less than 2.5%;
- minimum amount of impurities, for example, H_2S.

Following is an example of crude oil speci cations:

- true vapor pressure (TVP) not exceeding 0.8 bar at highest loading temperature;
- base sediment and water (BS&W) less than 0.5% (v/v);
- salt content less than 200 mg/l.

5.2.2 Wellstream properties

The wellstream uids owing from the reservoir are broadly categorized using the following properties (Jahn et al., 1998):

- The gas to oil ratio (GOR), which is the ratio of the volume of gas to the volume of oil produced at standard temperature and pressure (25 °C, 1 atm).
- The oil gravity. It is measured in degrees API, which is de ned by the American Petroleum Institute as follows: API = $(141.5/\gamma)$ 131.5. Here, γ is the speci c gravity of the oil with respect to water.
- Fluid composition.

The six main categories of well uids are (Jahn et al., 1998):

Dry gas: Colorless natural gas that appears in the presence of little or no hydrocarbon liquids (condensate). The composition of dry gas is mainly C1 (methane), and the GOR is larger than 50,000 Sm^3/Sm^3.

Wet gas: Colorless natural gas with some clear liquids (condensate). Composition is mainly C1 (methane), but less than in dry gas. The GOR is larger than 3000 Sm^3/Sm^3, and the API of the condensate is typically 60 70. *Note*: When a gas is said to be *wet*, it refers to the NGL that condense from the gas, not the water content.

Gas-condensate: Colorless natural gas with signi cant amounts of clear liquids (condensate). C1 (methane) content is typically about 70 mole%. The GOR is larger than 700 Sm^3/Sm^3 and the API of the condensate is typically 50 70. The condensate is mainly composed of C3 (propane), C4 (butane), C5 (pentanes), and C6+ (heavier hydrocarbon fractions). The formation of condensate depends on pressure and temperature, and may be generated in the reservoir as well as formed when liquid drops out, or condenses, from a gas stream in pipelines or surface facilities.

Volatile oil: Brown oil with a GOR of 250 500 Sm^3/Sm^3. The gravity of such oil is typically 40 50 API.

Black oil: Black, highly viscous (high ow resistance) oil with a GOR less than 250 Sm^3/Sm^3. The gravity of such oil is typically less than 40 API.

Extra heavy crude: Black, highly viscous oil with a very low GOR. The gravity of such oil is typically less than 10 API, that is, heavier than water. Normally the crude must be diluted to make it ow. Naphtha is often used as a diluent.

Representative reservoir uid samples must be obtained as a basis for designing process facilities. The uid properties are measured in the eld either by bottom-hole uid samples or by taking samples from a test separator, depending on the reservoir characteristics. Bottom-hole sampling will only be successful in saturated reservoirs, that is, reservoirs whose pore structures are lled with oil. In other reservoirs, it is recommended to sample the uids in the separator. The most crucial issue is to obtain a representative and correct GOR, and to ensure that the well ow is stable while sampling.

Table 5.1 shows an example of molar composition of reservoir uids. The left column speci es the size of molecules, by their number of carbon atoms, whose fractions are speci ed in the other columns.

A high gas/oil ratio in a crude oil wellstream necessitates large processing facilities in order to process the oil to stable condition. In addition, the following uid and well properties in uence the design of the processing facilities:

- oil gravity;
- hydrocarbon composition;
- acid gas content (H_2S and CO_2);
- other contaminants presence (N_2, mercury, mercaptans, asphaltenes, wax, etc.);
- amount of formation water;
- emulsion and foam behavior;
- hydrate formation and scaling;
- sand production from the well;
- uid viscosity.

Table 5.1: Example of a molar composition of reservoir uids. (Note that the terms N+ in the columns represent the contributions to the molar composition of heavier hydrocarbon fracture components; CN and higher).

Component	Gas	Gas-condensate	Volatile oil	Black oil
N_2	0.3	0.71	1.67	0.67
CO_2	1.1	8.65	2.18	2.11
C1	90.0	70.86	60.51	34.93
C2	4.9	8.53	7.52	7.00
C3	1.9	4.95	4.74	7.82
C4	1.1	2.00	4.12	5.48
C5	0.4	0.81	2.97	3.80
C6	*6+:* 0.3	0.46	1.99	3.04
C7		0.61	2.45	4.39
C8		0.71	2.41	4.71
C9		0.39	1.69	3.21
C10		0.28	1.42	1.79
C11		0.20	1.02	1.72
C12		0.15	*12+:* 5.31	1.74
C13		0.11		1.74
C14		0.10		1.35
C15		0.07		1.34
C16		0.05		1.06
C17		*17+:* 0.37		1.02
C18				1.00
C19				0.90
C20				*20+:* 9.18

Of particular importance is any presence of H_2S, CO_2, other detrimental gases, rare metals, or radioactive substances. Corrosive substances may necessitate special corrosion-resistant construction material and the need for speci c processes to be included.

The GOR, amounts of condensate, the oil grade (heavy, light, etc.), and water content form the basis for process requirements. The presence of water necessitates gas drying. Oil reservoirs may require facilities for handling large amounts of water.

For oil reservoirs containing substantial amounts of associated gas, gas disposal may present a challenge. Historically it has been common to are off such gas, particularly in areas that lack gas pipeline connection to any large market for gas. The practice of aring is increasingly regarded as unacceptable on environmental grounds, and also because it is a waste of a resource that has become increasingly valuable. Where no gas offtake option is currently feasible, gas re-injection should be considered, either into the oil reservoirs where it can also be useful for pressure maintenance (see Chapter 2) or into a separate structure. Injected gas can usually be recovered later to a large extent. Gas re-injection may require large compression capacity to be installed.

5.3 A functional overview of oil and gas processing

A generalized block diagram for an oil and sales gas processing facility is shown in Figure 5.1. Some of the blocks shown may be interchangeable. For example, the gas may have to be dehydrated (removal of water to acceptable limits) *before* it is compressed. Also, the gas will have to be recompressed after it has been treated. Additional components may be necessary, such as water, gas, and CO_2 injection facilities, desalting facilities, desanding facilities, etc.

5.3.1 Heating

It may be necessary to heat the wellstream, especially if it contains oil. Following are reasons for heating the wellstream:

- minimize wax, scale, and asphaltene deposition;
- break emulsions and foams;

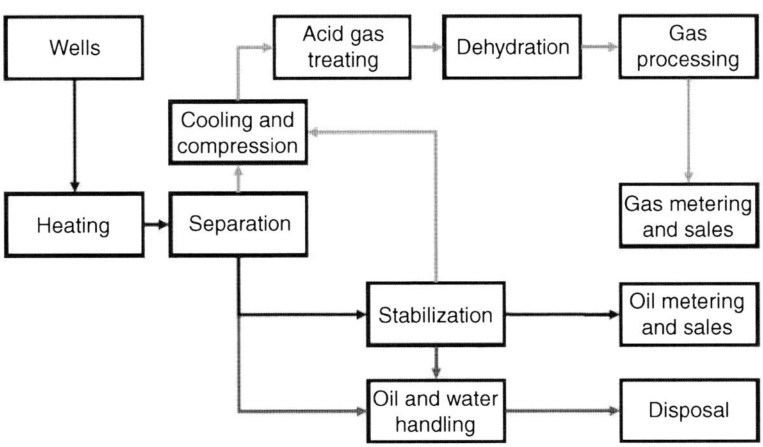

Figure 5.1: Block diagram of a process facility (illustration: B. Hurle).

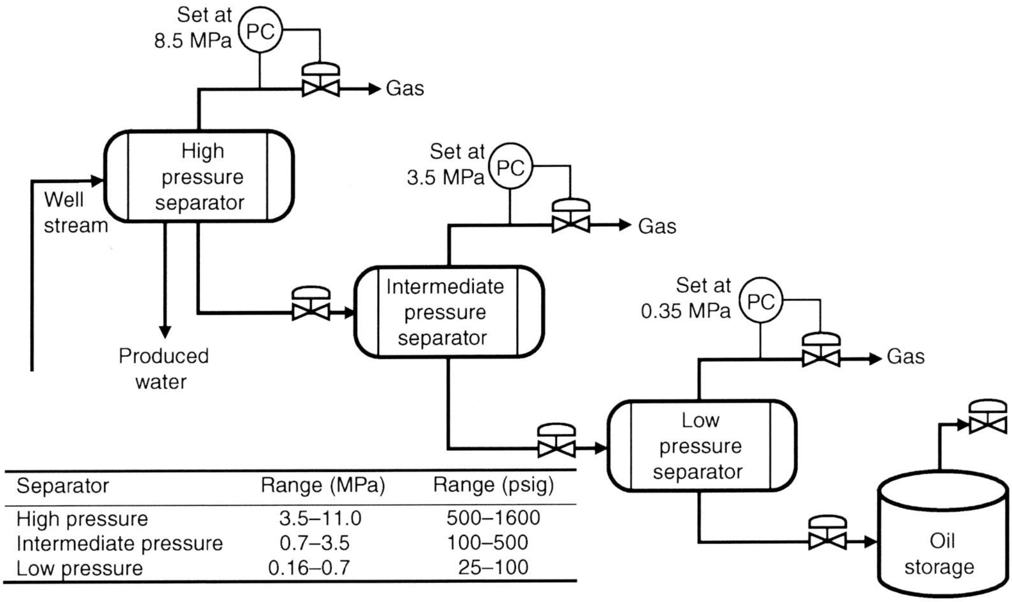

Separator	Range (MPa)	Range (psig)
High pressure	3.5–11.0	500–1600
Intermediate pressure	0.7–3.5	100–500
Low pressure	0.16–0.7	25–100

Figure 5.2: Multistage separation. *Note*: One standard atmosphere pressure is 1 atm = 0 psig (gauge) = 14.7 psia (absolute) = 101.3 kPa = 0.1013 MPa = 1.013 bar.

- avoid hydrate formation;
- reduce the viscosity of the oil (make it ow more readily);
- separate gas from solution in the oil.

5.3.2 Separation

The purpose of the separator is to get gas out of solution, obtain a more moderate pressure and temperature, and remove free formation water from the oil/condensate stream. Often, the separation process is accomplished in several distinct stages. When moving from one stage to the other, the stream is said to *flash*. It then changes its PVT properties. The purposes of multi-stage separation (Figure 5.2), as opposed to single-stage separation, are to obtain better liquid recovery (i.e., condensation of heavy gas components) and to use smaller units.

5.3.3 Oil and condensate stabilization

Stabilization of condensate or oil refers to the adjustment of the vapor pressure and composition of the liquid to meet transport properties. Even though the oil/condensate phase has been largely separated from the gas and water phases, it still contains gas and water in solution. By dilution, the volatile components (the gas in solution) may cause the heavier components in the oil to vaporize, which is undesirable. To restore equilibrium and obtain a vapor pressure of less than 0.8 atm, some of the volatile components must be removed. This is often done using a small fractionation column, as indicated in Figure 5.3. Fractionation is a separation process

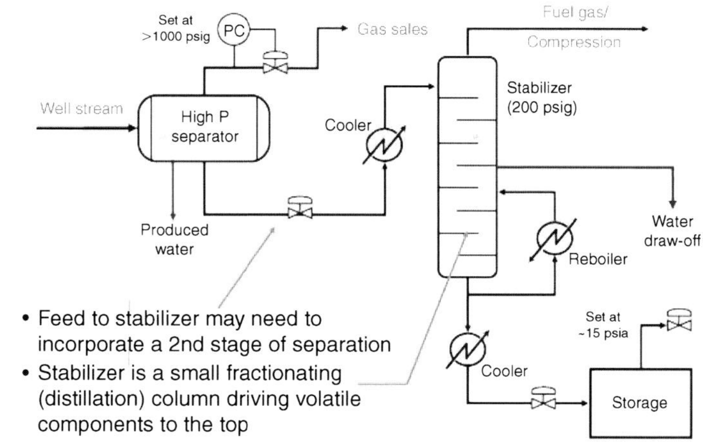

Figure 5.3: Condensate stabilization column (Hurle, 2006).

in which a mixture is divided into a number of smaller quantities in which the composition changes due to a gradient, usually a concentration gradient or temperature gradient. In the oil and gas industry, the most widely form of fractionation used is *fractional distillation*, in which components are separated due to differences in boiling point.

As indicated in Error! Reference source not found., the bottom of the stabilization column is hot, and the top of the column is cold. Only heavy components will stay in the liquid phase at the bottom, and only volatile components (gas) will stay in the gas phase at the top. This makes it possible to extract the heavy components at the bottom of the column and the gas at the top.

5.3.4 Water handling

Hydrocarbon production involves complying with water quality standards that are designed to meet appropriate health and environmental standards. Petroleum production generates water in various chemical compositions, and as a result, requires the development of customized water treating systems.

Handling the produced water within environmental limits requires careful consideration of the following produced water contaminants:

- hydrocarbons dissolved in the water;
- scaling in water due to high concentrations of carbonate reacting to form solid precipitates.

The amount of water contained in wellstreams varies between ! elds, and may also vary over the life of the ! eld, with increasing water content occurring in the later stages of ! eld production. Initial water content in an oil well may be close to zero, but may exceed 90% as the well approaches abandonment. Water handling is therefore a key component of production operations particularly in aging ! elds, and an important constraint on the ability to achieve a high ultimate oil recovery.

The formation water separated from the wellstream may be re-injected into the reservoir, or treated and disposed of. The water may have to be treated in either case. The most basic

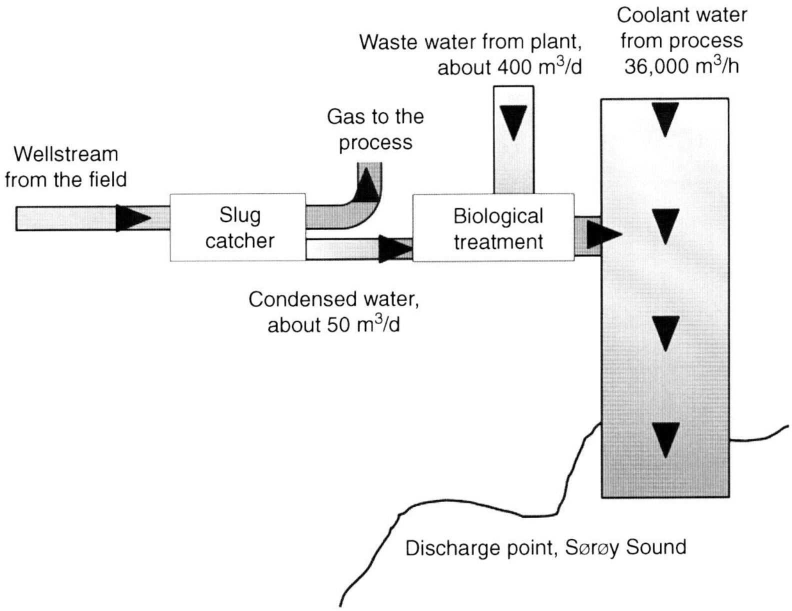

Figure 5.4: The Sn! hvit water treatment concept.

form of deoiling facility is a skimming tank. It may reduce the oil concentration in the water to 200 parts per million (ppm). However, to meet disposal standards, more sophisticated methods may be necessary. Gas ! otation and hydrocyclone processes are two techniques that can reduce the oil concentration to 40 ppm (Jahn et al., 1998).

Biological treatment methods are also available. One such method is used at the Melk! ya LNG plant for processing the gas/condensate wellstream from the Sn! hvit ! eld in Norway, as indicated in Figure 5.4.

Veil et al. (2004) have produced a useful reference publication for water treatment requirements directed at the US petroleum sector, but relevant also elsewhere.

5.3.5 Cooling and compression

The gas that has been ! ashed off during separation and the gas that has been forced out of the solution during stabilization are cooled, and may also have to be compressed prior to further processing. The amounts of cooling and compression required depend on the pressure and temperature of the gas stream. It serves the following purposes:

- increase the gas density to maximize throughput and minimize equipment size;
- maximize ef! ciency;
- meet the inlet pressure and temperature speci! cations of the subsequent processing equipment;
- knock out water and heavy hydrocarbons carried forward in the gas stream.

The gases coming from the different separators have different pressures and temperatures. Where the pressure of the feed gas is too low, a compressor will have to be used, and simple refrigeration or heat exchange with colder streams is used to cool the gas.

5.3.6 Acid gas treatment

The main acid gases in wellstreams are CO_2 and H_2S. They need to be removed from the gas stream for the following reasons:

- toxicity (H_2S);
- corrosiveness;
- catalytic properties;
- thermodynamic properties;
- environmental concerns.

H_2S is a highly toxic and highly corrosive compound. Any high concentrations of it will require special corrosion-resistant steel pipes. H_2S acts as a catalyst for certain undesirable chemical reactions.

CO_2 is corrosive if contacted by water, forming carbonic acid. If cooled to −78.5 !C, a temperature that is commonly reached in cryogenic gas fractionation and LNG production, it solidi! es and forms *dry ice*. Dry ice can, like hydrates, clog up pipes and equipment such as compressors and turbo-expanders. Furthermore, CO_2 is a greenhouse gas and may have to be removed for environmental reasons.

H_2S and CO_2 may be removed by absorption, adsorption, or gas permeation processes. *Absorption* is a process in which the gas stream is fed into the bottom of a column, where it travels upwards and is allowed intimate contact with a solvent, usually amines or potassium carbonate (K_2CO_3) for H_2S and CO_2 removal, as indicated in Figure 5.5. Acid gas removal by absorption does not remove other contaminants containing sulfur such as mercaptans (R-SH), COS, SO_2, and CS_2.

Acid gas removal by *adsorption* is a process in which the gas is fed through a highly porous material, usually a molecular sieve. In gas treatment by adsorption, the unwanted contaminants are trapped in the tiny pore spaces in the sieve, and the sieve will have to be heated or swept with an inert or clean gas to get them out again. This process is called *regeneration*. Adsorption towers are usually used when very high gas purity is required. Molecular sieves will remove CO_2, H_2S, SO_2, NH_3, COS, mercaptans, and water. However, for molecular sieve towers to be used, the water content of the feed gas has to be relatively low. Also, COS and CS_2 will deteriorate the sieves. For this reason, adsorption towers may have to be used in conjunction with absorption columns. An adsorption process layout is shown in Figure 5.6.

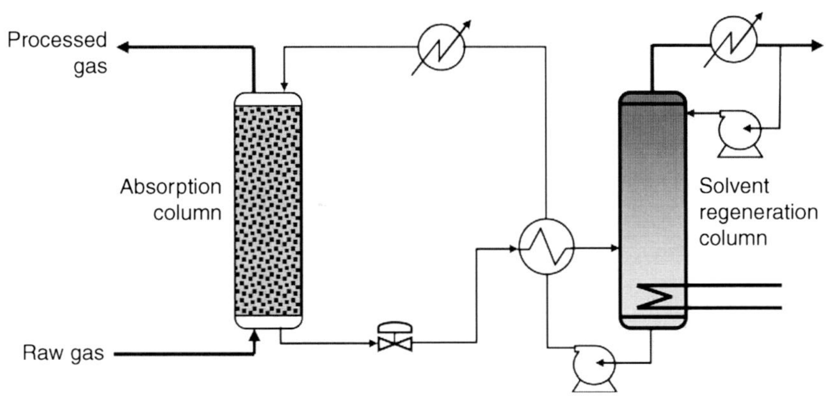

Figure 5.5: An absorption column (illustration: adapted from Hurle, 2006, with permission).

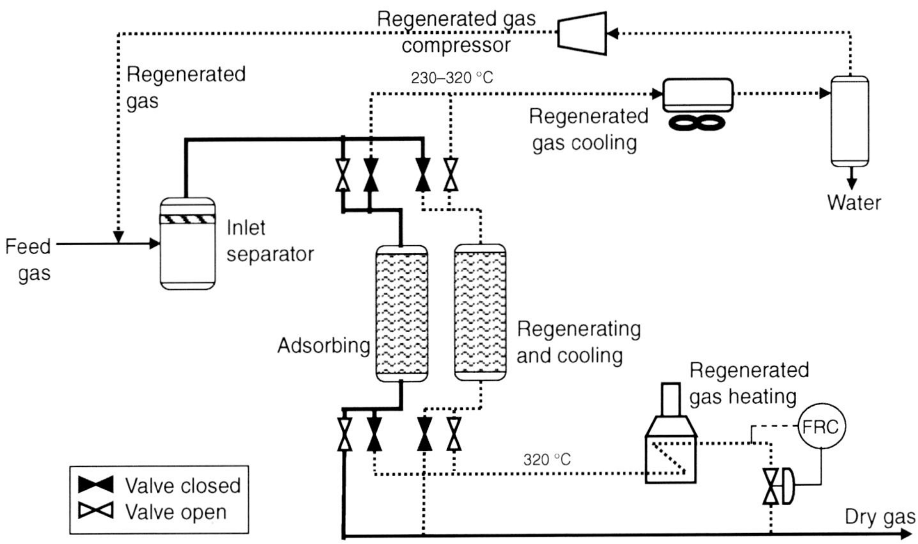

Figure 5.6: An adsorption process layout (illustration: adapted from Hurle, 2006, with permission).

Another technology that will remove CO_2 from the gas stream is *gas permeation*, in which a membrane allows some components to pass through while others are rejected. In the correct con! guration, gas permeation can be very selective, and is proven to be a viable alternative to water removal by absorption. Gas permeation will also reduce the H_2S content of the gas, but a ! nal speci! cation cannot be guaranteed. Gas permeation is a promising technology that is still under development. However, technical problems have to be overcome for the method to gain wider acceptance.

5.3.6.1 CO_2 injection at the Sleipner and Snøhvit fields (Norway)

Rather than being released into the atmosphere, the CO_2 produced with the gas from the Sleipner ! eld outside western Norway is injected into a separate geological structure near the main ! eld. One million tons of CO_2 is injected annually.

At the Sn! hvit ! eld in northern Norway, the CO_2 is stored 2600 m beneath the seabed at the edge of the reservoir. A total of 700,000 tons will be stored annually, which is equivalent to the emissions from 280,000 cars.

The natural gas from the Sn! hvit ! eld contains 5! 8% CO_2. The greenhouse gas follows the wellstream from the ! eld to the land terminal, where the CO_2 is removed and piped back to the ! eld.

The process of separating the CO_2 from the natural gas entails combining amine and natural gas in a tank at high pressure and moderate temperature. The amine binds itself to the CO_2 and separates out at the bottom of the tank, and is transferred to a new tank where the pressure is lower and the temperature higher. The greenhouse gas is removed and sent from the tank to the CO_2 injection system. Gas production is expected to reach 6.9 billion Sm^3 per year, and is exported as LNG.

The CO_2 is injected into a sandstone formation, which is between 45 m and 75 m thick and lies somewhat deeper than the gas formations. A sealing formation above the sandstone will ensure that the CO_2 does not leak out.

5.3.7 Dehydration

The gas that is ! ashed off in the separators will contain considerable amounts of water vapor. There are several reasons for removing this water from the stream:

- Water will react with acid gases such as H_2S, CO_2, and SO_2, causing corrosion.
- Water vapor may condense or freeze at low temperatures, clogging pipes and damaging equipment such as compressors and turbo-expanders.
- Water may react with the light alkanes (methane, ethane, etc.) to form hydrates, resulting in similar problems as those for ice and dry ice.
- Water has no commercial value as a component of gas.

The most common method for removing water from a gas stream is by absorption. The most widely used solvent is tetraethylene glycol (TEG), often just referred to as !glycol.! However, there are other commercially available glycols. In a glycol absorption process, water-rich gas is fed into the bottom of a column. Regenerated glycol is fed into the top. The gas percolates up the column and meets progressively richer glycol.

There are temperature and pressure restrictions applicable to glycol columns. The temperature in the column should be limited to about 35 !C to limit the glycol losses to the gas stream, and to improve the water absorption capacity of the glycol. This will also limit the amount of water present in the vapor phase. The operating pressure of the column should be at least 150 bar. Hence, there is a need for cooling and compression of the gas stream. Glycol absorption technology can reduce the water content of a gas stream to about 45 ppm.

When water content speci! cations are very restrictive, as is the case in LNG production, dehydration by adsorption may be necessary. Molecular sieves are most commonly used, and these will remove several contaminants simultaneously. However, the compound for which the molecular sieves have the highest af! nity is water, and water will thus displace all other contaminants in the sieve. The af! nity ranking is as follows: water > mercaptans > H_2S > CO_2. Dehydration by adsorption may reduce the water content to about 0.03 ppm, and is far superior to glycol absorption. It is, however, much more expensive.

As mentioned in the preceding section, water may also be removed by gas permeation. Cooling the gas will also knock out water by condensation; this is done prior to absorption or adsorption.

5.3.8 Other gas processing requirements

There may be a need to remove further contaminants from the gas stream, like mercury (if present). Also, further re! nement of the gas stream may be required for the gas to obtain sales speci! cations, as the gas resulting from the previously mentioned processes is still likely to contain signi! cant amounts of NGL.

NGL recovery is usually done by cryogenic (very low temperature) fractionation, in multistage distillation. The gas stream is fed through several fractionation columns, where one or more components may be removed at a time. The principle used in fractionating columns is that different hydrocarbon gases will condense at different temperatures, and may thus be separated. The most volatile alkanes have the following approximate boiling points at the pressure of 1 atm:

C1 (methane): −161.5 !C;
C2 (ethane): −88.6 !C;
C3 (propane): −42.1 !C;
i-C4 (*i*-butane): −11.8 !C;

n-C4 (*n*-butane): −0.5 !C;
i-C5 (*i*-pentane): 27.8 !C;
n-C5 (*n*-pentane): 36.1 !C.

The methane may be exported as sales gas in pipelines, or lique! ed and sold as LNG. It may be necessary to keep some heavier components in the sales gas to boost its heating value to meet the quality speci! cations agreed upon with the gas distributor.

Considerable amounts of energy are required to liquefy and recompress natural gas, and several methods may be employed. Where the pressure of the gas stream is high, it may be cooled by throttling it through a Joule!Thomson (JT) expansion valve or expanding it using a turbo-expander. These are very expensive units that may be described as compressors working in reverse. Where a high-pressure gas stream is not available, simple refrigeration is needed, which requires large amounts of energy.

5.3.9 Natural gas processing and terminal ling at Kårstø

The K!rst! natural gas processing and terminal plant, in southern Norway, is the largest of its kind in Europe.

The K!rst! plant is connected by two rich gas pipelines from offshore, capable of delivering more than 100 million Sm^3 per day. There is also a liquid line carrying about 20,000 tons of unstable condensate daily from the Sleipner ! eld (see above).

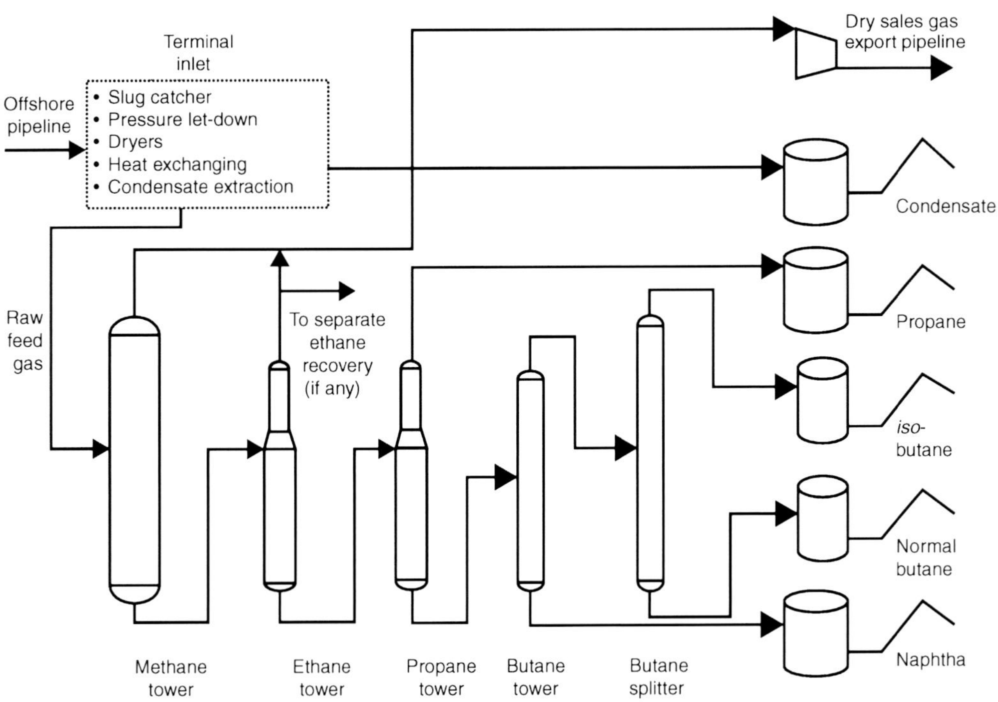

Figure 5.7: Simpli! ed natural gas processing ! ow diagram (illustration: K. Traa).

Processing of the incoming product streams into sales gas, NGL products, naphtha, and stabilized condensate is carried out as shown in the simpli! ed processing diagram in Figure 5.7. Following pressure let-down, ! nal drying, and heating, the rich gas mixture is fed into a series of fractionation columns, where the light ends (starting with methane) are successively extracted at the top of the columns, while the heavier components are collected at the bottom. This process is governed by successive pressure reductions and temperature differentials, allowing heavier products to be extracted in liquid phase at the bottom of the fractionation column, while lighter products can be extracted as gases higher up in the column.

The sales gas produced is exported by large-diameter pipelines, while the liquid products are exported by smaller and larger ships, of which there may be more than 700 calls annually.

5.4 Summary

The amount of processing required for a wellstream largely depends on the wellstream ! uid properties and the end-product speci! cations. Other factors that need to be taken into consideration are transport requirements, reservoir injection requirements, storage requirements, and operational requirements.

The properties that specify the end product are different for oil and gas. In general, the sales gas speci! cations are the most severe.

The need to process a wellstream is also determined by its properties, notably the GOR, the API degree and viscosity of the oil or condensate, the composition of hydrocarbons, the acid gas content (H_2S and CO_2), the amount of other contaminants, the amount of formation water, the emulsion and foam behavior of the wellstream, and the amount of scaling and hydrate formation in pipelines and equipment.

The main functional blocks at a facility that processes the wellstream are heating, separation, stabilization, cooling, compression, acid gas treatment, dehydration, and water handling. Further re! nement, gas liquefaction, and other contaminants! removal may also be necessary.

References

Hurle, B. 2006. *Lecture Notes from the Course RESR3530 Process Modules*. The University of Western Australia, Perth, Australia.

Jahn, F., Cook, M. & Graham, M. 1998. *Hydrocarbon Exploration and Production*. Elsevier B.V., Amsterdam.

Veil, J.A., Puder, M.G., Elcock, D. & Redweik, R.J. 2004. *A White Paper describing Produced Water from Production of Crude Oil, Natural Gas, and Coal Bed Methane*. Prepared for the U.S. Department of Energy, National Energy Technology Laboratory, January, http://www.ead.anl.gov/pub/doc/ProducedWatersWP0401.pdf.

Hydrocarbon offtake

Oil and gas must be brought to buyers and users in order to be valuable. This chapter reviews options for transferring them from producing installations. It also reviews the ways in which oil, gas, and the natural gas liquids (NGLs) may be sold. Both issues have implications for the processing operations as reviewed in the previous chapter.

6.1 Alternatives for offtake: pipelines, ships, or local conversion

Pipelines and ships are the two main methods for transporting hydrocarbons to users. Both may apply to oil as well as gas. There are well-developed methods for constructing pipelines both onshore (Figure 6.1) and offshore. A third alternative offtake solution, particularly for natural gas, is to convert it to another product locally rather than exporting it.

6.1.1 Distance and volume as a determinants of offtake solution

More than 50% of all crude oil produced globally is moved by pipeline only from the producing field to a refinery. The remaining part is moved by ship (or, in some cases, on railway or

Figure 6.1: Construction of a large gas pipeline in Germany. The pipeline was buried before commissioning. Photo from Wingas.

trucks) for at least part of the journey. The corresponding figure for natural gas is more than 90% by pipeline only. Large, integrated pipeline systems for oil as well as natural gas are found primarily in North America and in Europe including Russia. Figure 6.2 shows a map of offshore pipeline systems used for Norwegian oil and gas exports. Ship transportation is used

Figure 6.2: Pipeline systems for transporting Norwegian oil and gas (illustration: Norwegian Petroleum Directorate).

for oil exports from the Middle East, Africa, South America, and Australia, as well as most Norwegian crude oil. Ship transportation of natural gas as liquefied natural gas (LNG) is most common for trade from Asian producers into Japan and Korea, but is increasingly used also at other export and import locations.

Different cost structures of pipeline and ship transportation tend to make them suited to different circumstances. Ship transportation is competitive for very long distances where pipeline construction would be immensely expensive; therefore, pipeline solutions of more than 2000 km are uncommon except where necessitated by the absence of suitable ocean access. Smaller distances to market and large volumes of product tend to favor pipeline over ships as an export solution.

Oil and gas pipelines require large up-front capital expenditures, but relatively small operating and maintenance costs later. They are long-lived, upward of 50 years for properly constructed and maintained lines. Pipelines are however rather inflexible as a transport solution: once installed, there are very limited options for modifying or redeploying a pipeline to another use. The capacity of a pipeline in terms of volumes of oil or gas transported is a function of its diameter as well as the pumping (compression) power applied. For a given pipeline diameter, it can often be possible to increase transport capacity by installing more oil pumping or gas compression power, within the limits determined by the structural strength of the pipeline.

Ship transportation is inherently more flexible than pipelines: whereas a pipeline may link an oil field physically to one or a few potential buyers (refineries), oil on a ship can reach the majority of the world's more than 500 refineries. Most oil tankers are not constructed for any specific source or destination of trade, but can be chartered and redeployed world wide as needed. An exception in this regard are offshore loading vessels such as shown in Figure 6.3, due to their specialized equipment for such loading operations. In contrast to most conventional oil tankers, the ships carrying natural gas (LNG) have usually been constructed for one particular trade, that is, traveling always between the same two ports, but some such ships have since 2000 been increasingly deployed in more flexible trading patterns.

Figure 6.3: A crude oil tanker equipped for offshore loading in the North Sea. Photo: Knutsen OAS Shipping.

For natural gas, the transportation by ship requires liquefaction by cooling the gas to −163 °C (LNG). The liquefaction reduces the gas to 1/600 of the gas phase volume so as to make ship transportation feasible. This is a costly process with regard to the required ships as well as the terminals at both ends, making ship transportation of natural gas several times as costly per energy unit as for crude oil. The high cost of export solutions for natural gas has resulted in large amounts of associated gas at remote locations being flared. Global LNG trade however has increased strongly after year 2000. With the high prices of oil as well as gas reached after 2005, expensive solutions for moving remote gas to markets either by pipeline or as LNG can increasingly be justified on economic grounds.

6.1.2 Implications for the natural gas liquids

The choice of pipeline or ships as offtake solution has implications also for process configuration. Additional to crude oil and/or natural gas processed to sales specifications, many fields also yield NGLs.

The relatively simple processing that can be done on an offshore installation usually allows for only two export streams, that is, gas and liquid. The NGLs may then be included with either of these, resulting in different limitations for offtake solutions. A liquid (oil) stream containing large portions of NGLs usually cannot be exported on ships, because the vapor pressure would be too high. Pipeline export is however possible for such a liquid stream, in which case the NGLs are recovered at an onshore terminal or at oil refineries as LPG and naphtha.

Alternatively, the NGLs may be included in the gas export stream, leaving a stable crude oil that can be moved on ships. This results in a "rich" gas stream that has too much NGL both for common sales specifications and for transportation by ship as LNG. Such "rich" gas (or field gas) is therefore commonly transported to gas processing plants which separate the NGLs from the dry gas and provide for separate exports of each. The Kårstø gas plant in western Norway receives "rich" gas from the oil fields in the northern North Sea, and separates the ethane, propane, butanes, and condensate for export by ships (Figure 6.4). This leaves a dry natural gas which meets European sales specifications and is exported by pipeline to the European continent. LNG liquefaction (not present at the Kårstø plant) always requires removal of the NGLs and impurities from the gas, as the heavier gas components would cause problems at the low temperatures involved.

6.1.3 Local use and conversion as an alternative offtake solution for gas

Since long-distance transportation of gas tends to be very costly, it is often relevant to consider using the gas locally as an alternative. Distribution of gas to local industries, businesses, and homes that may use it as a fuel can be considered, but would in many cases not provide any large outlet for gas. Some countries that have domestic gas resources use it as a fuel for road vehicles, for which purpose the gas is compressed for storage in vehicle fuel tanks as compressed natural gas (CNG; not to be confused with LPG, i.e., propane and butane, which are also used for similar purposes). Argentina, Brazil, and Pakistan have implemented this on a large scale.

Electricity can be generated from natural gas on a small scale in reciprocating engines (resembling diesel generators), and on a larger scale in gas turbines or in the more sophisticated and efficient combined cycle power plants. The viability of producing electricity from gas depends on the local demand and transmission grid for electricity, as long-distance

Figure 6.4: The Kårstø gas terminal. Photo: Statoil/Dag Myrestrand.

transport of large amounts of electricity tends to be less cost efficient than corresponding long-distance transport of natural gas.

Natural gas can also be converted chemically to certain other products, the most common being ammonia and its derivatives (notably urea and nitrous fertilizers). Liquid fuels (especially diesel) can be made synthetically from coal as well as natural gas, referred to as gas-to-liquids (GTL) technology and related to the Fischer–Tropsch synthesis. Such plants currently operate in South Africa, Qatar, and Malaysia. Several new GTL projects have been announced after year 2000, but a number of them have since been discontinued due to concerns over high costs.

Before embarking on any industrial gas conversion project, it is advisable to assess very carefully its industrial and competitive viability, bearing in mind that the products will have to compete in markets that have not always provided much profit for producers.

6.2 Pipeline design considerations

Pipeline design needs to be determined by anticipated product specification, pressure conditions, and flow rates. Pressure conditions include the internal pipeline pressure and, with particular regard to deepwater pipelines, external pressure.

For oil pipelines, the physical characteristics of the oil to be moved result partly from the composition of hydrocarbons in the reservoir, and partly from the processing that is done upstream of the pipeline. Potential flow problems associated with waxy or heavy viscous oils may need to be dealt with either by upstream processing or by measures along the pipeline. The latter may include heating, insulation, use of admixtures, and frequent pigging.

In pipelines carrying "rich" gas (i.e., natural gas with large contents of condensate, butane, and propane), pressure must be managed so as to prevent liquids from separating from the gas in the pipeline. Inhibitors such as methanol or glycol may be required to avoid hydrates.

To properly design a pipeline, expertise within the following subjects will be required:

- process design;
- pipeline system design as well as resulting compression needs, and determining pipeline diameter and internal pressure;
- aspects related to external vs. internal design pressure, determining the steel-wall thickness;
- pipeline project execution, including expertise in pipe laying and operations;
- fluid composition, determining the steel quality and the possible needs for internal coating.

The following engineering aspects must be incorporated in the pipeline design process:

- pipeline routing to preferably avoid uneven terrain or sea-bottom conditions as well as any particular environmental or man-made obstacles;
- evaluation of possible solutions for sea-bottom free-span conditions for offshore pipelines;
- design of external corrosion coating;
- design of compressor stations and measurement stations as well as pig (scrape) launchers and receivers;
- detailed account of the connections between pipelines and platforms or terminals, taking into consideration the expansion of pipelines carrying warm fluid;
- for offshore pipelines, checking the steel thickness to see whether the pipeline can withstand the external seawater pressure during and after installation;
- design of weight coating to ensure bottom stability of a pipeline in water;
- deciding the depth below surface of an onshore pipeline to minimize future conflict with land use (for an offshore pipeline, consider trenching to avoid damage from anchors, fishing gear, etc.);
- planning for required safety zones and restrictions on activities near the pipeline;
- detailed account of critical crossings, shore approach, as well as the terminal end of the pipeline;
- optimization of life cycle costs, that is, initial investment in relation to duration, future maintenance, and operating costs for the pipeline.

For onshore pipelines in particular, pipeline construction often requires obtaining the consent of (private) owners of the land, as well as the relevant authorities. The appropriate approach to this varies with local conditions and legislation. Often the pipeline owner can obtain the right to install the pipeline on someone's land while not acquiring ownership to the land. The use of the land for agricultural and other purposes can go on, but with certain stipulated limitations providing for safety, protection of the pipeline's integrity, access for the pipeline operator for inspection purposes, etc.

For relatively short pipelines, the use of flexible lines can be considered, depending also on the maximum pressure and temperature of the fluid that may be transported. Figure 6.5 shows an example of flexible pipe design. In other circumstances, pipelines for local gas distribution at less than 10 bar pressure can be made entirely of polyethylene.

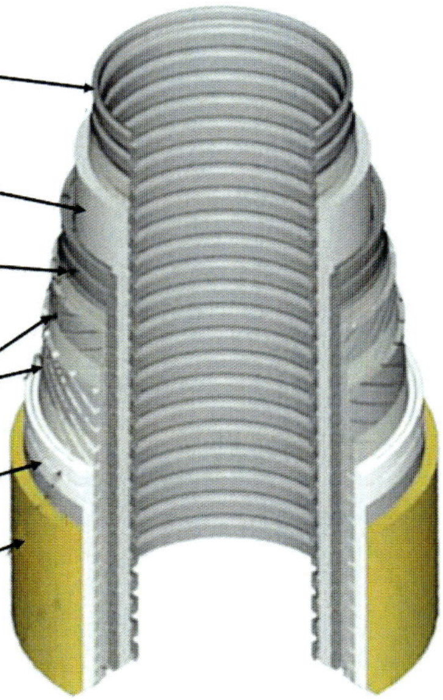

Figure 6.5: Flexible pipeline (illustration: FlexiRiser, www.flexirisertest.com).

6.3 Measurement of hydrocarbons

Measurement of hydrocarbon liquids and gases is an essential part of the whole hydrocarbon value chain, from wellstream to the final sales products. It serves the following purposes:

- wellstream monitoring, that is, to monitor the performance of a reservoir;
- allocation metering, for determination of ownership and for calculation of user tariffs;
- determination of quantities related to calculation of fiscal royalties and taxes;
- custody transfer of products and metering of sales volumes/quantities.

Metering of hydrocarbon liquids and gases is carried out according to several methods, any of which must meet the requirements of international standards (such as ISO and AGA), as well as national regulations, such as the Norwegian Petroleum Directorate's regulations relating to the fiscal metering of oil and gas. The Norwegian Petroleum Directorate's fiscal measurements regulations (2001) can be found at www.npd.no.

Common measurement methods are the following:

- Orifice plate meter: Measuring the pressure drop through an orifice plate, that is, the flow rate is a function of the square root of the pressure difference across the orifice plate.
- Turbine meter: Measuring the flow velocity, that is, the force of the flowing fluid turns a bladed rotor in a turbine meter.
- Ultrasonic meter (Figure 6.6): Measuring the mass flow, whereby the difference in the transit time measured when alternately transmitting and receiving a burst of sound energy between two transducers is directly and exactly related to the velocity of the fluid in the pipe.

Illustration examples of the metering systems can be found at www.emersonprocess.com (orifice meter), www.shenitech.com (ultrasonic flow meter), and www.fmctechnologies.com (turbine flow meter).

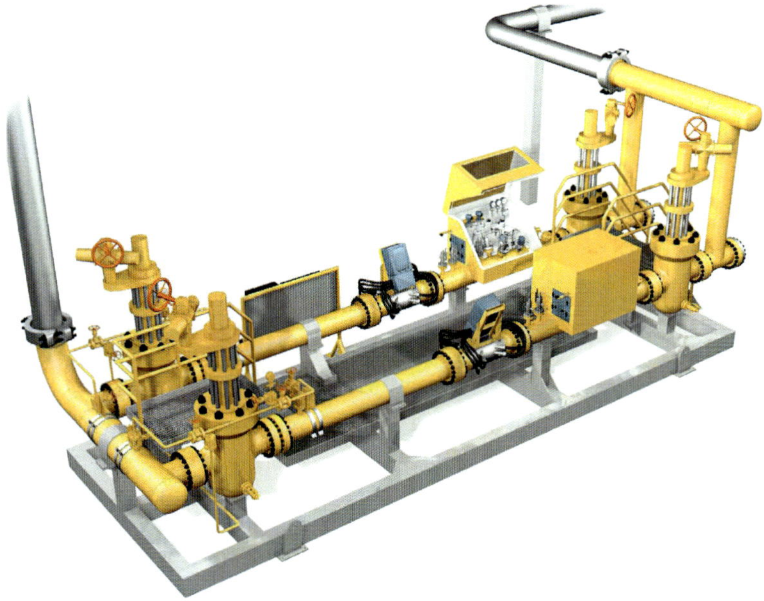

Figure 6.6: Ultrasonic gas metering station (illustration: FMC Technologies).

Besides metering of flow (i.e., quantity), the quality (composition) of hydrocarbons must also be determined. For crude oil, the main quality parameters are vapor pressure, water content, salt, and bottom sediments, which are usually determined from laboratory samples.

Gas quality measurements including density, composition, calorific value, and Wobbe Index are normally determined on-line, by instruments connected to a meter station. A gas chromatograph is used to determine the molecular composition of gas. On this basis, the calorific value of natural gas is calculated. The calorific value is particularly important for delivery under sales contracts, since the price to be invoiced usually refers to the energy content of the gas.

All metering facilities and related quality measurements must be calibrated and checked regularly, in accordance with recognized codes and standards, to keep all measurements within prescribed tolerances.

6.4 Terminals and processing plants

The term *terminals* in the oil and gas industry usually refers to facilities at the ocean shore which provide for the receipt of oil and/or gas from producing fields, in some cases storage, and onward transfer by either pipeline, ships, or land. Terminals often also include some kind of processing of hydrocarbons. Following is a list of types of terminals related to oil and gas production according to their different purposes:

- Treatment of gas to sales gas specification with extraction of LPG and condensate products (example: the Kårstø terminal; see Figure 6.4).
- As above but with facilities to produce LNG for ship transport to markets (example: several plants in Qatar; the Norwegian Snøhvit terminal at Melkøya).

- Treatment of oil or condensate with extraction of different products (the Teesside terminal in the United Kingdom as well as one part of the Kårstø terminal).
- Receiving terminal for sales gas with measurement of product flow and with receiving station for pigging (pipeline scraping) equipment (e.g., the Emden gas terminal in Germany, the Zeebrugge gas terminal in Belgium, or the Dunkerque terminal in France). The Emden terminal is also connected to salt cavern storage for gas, serving as a buffer for the gas buyers should the offshore gas production have to stop for any reasons.
- LNG-receiving terminals (example: the Isle of Grain terminal in England).
- Terminal for treatment of multiphase flow, that is, processing of the fluid with separation of water and condensate, and processing of the gas to sales gas specification (e.g., the Troll terminal onshore at Kollsnes in Øygarden at the western coast of Norway).
- Storage and shipment of crude oil (Sture terminal, near the Kollsnes terminal).
- Oil terminal associated with a refinery. In addition to being an oil refinery, the Mongstad site includes a large facility for storage and transshipment of crude oil (Figure 6.7).

Offshore storage tanks, concrete platforms, and production ships can also provide terminal functions.

The selection of the process to be carried out at the onshore terminal must be based on a cost optimization analysis where the costs of carrying out the processing at a terminal onshore

Figure 6.7: The Mongstad oil refinery and storage terminal. Photo: Statoil/H.M. Valderhaug.

must be compared with the costs of offshore processing. The costs of offshore processing are normally very high.

A terminal development project is in itself a major development project and must be treated as such. Following are important considerations for selecting a site for a terminal project:

- safety for personnel living in the vicinity of the facility, as well as safety for personnel working at the terminal;
- area requirement and possible conflicts with other industries and possible conservation requirements;
- feasibility of pipeline shore approach;
- evaluation of the distance and means for transport of products to the market;
- geotechnical conditions at the site;
- maximum use of infrastructure (e.g., water and electric power);
- availability of water for cooling;
- transport possibilities for the construction phase (boat, railway, or road).

Processes must be selected to provide environmentally acceptable products. The concept development phase of a terminal project must be executed in an orderly manner to facilitate accurate bidding and the possibility to carry out the construction project with a minimum of design and contract deviations.

6.5 Ownership, third-party use, and quality adjustment

Many pipelines carry oil or gas belonging to several parties, originating from several fields having different ownership.

There are strong economic reasons for this: where a new field is to be brought on stream near an existing pipeline, there can be large cost savings obtained from utilizing any free capacity in the existing pipeline rather than building a second one. There tends to be large economies of scale in pipeline construction, that is, by selecting a larger pipe diameter than what would be required for the initial transport needs, it can be possible to obtain large extra capacity at modest extra cost.

Long-distance transmission pipelines for oil and especially for gas regularly carry product from a multitude of origins. The European Union, the USA, and other countries have legislation which obliges owners of gas pipelines to offer their capacity to others (*mandatory third-party access*). Their main motive for this is to create competition between gas suppliers in the interest of consumers, but some producer nations (such as Norway) pursue similar policies for upstream pipelines in order to enable the optimization of the overall transport system. Some pipelines are owned and operated by a company that has been set up for that specific purpose.

With such multiparty use of pipelines, the ownership of a pipeline does not coincide with ownership of the product inside it. This raises a number of issues which must be carefully regulated by agreements between the parties involved. This is usually resolved with separate agreements for the joint ownership of a pipeline system (often as a partnership between several oil firms) and for transporting oil or gas in the system.

Transporters (i.e., owners of product in the pipeline) pay a tariff to the pipeline owner, from which the pipeline owner may recover operating costs and capital expenditures with a reasonable return on investment. The tariff may have several components, such as one charge for the amount of capacity reserved and another charge for the amount of product actually transported. Norway's petroleum authorities have a policy that the tariffs should allow for only a modest return on investment, based on a principle that profits in oil and gas production should accrue at the producing field, not the pipeline.

A pipeline system needs procedures for the allocation of capacity, especially when there is a possibility from time to time that transporters in sum would want to put more products down the pipeline than capacity allows. This requires rules for prioritizing capacity rights, capacity booking, and nominations on annual and daily basis. These are formal communications between the pipeline operator and transporters concerning transport volumes.

Certain pipelines carrying liquids, notably refined oil products, can operate in batch mode, that is, they are used for several different products which are sent down the pipeline one at a time so as not to blend them. Such batch operations would be impractical in most cases where crude oil or natural gas needs to be transported. Pipelines are often used to transport oil or gas from different sources having different quality, which are commingled in the pipeline. The product that a transporter can recover at the downstream end of a pipeline is therefore not the same product that he entered at the other end, but rather a blend of all products in the pipeline. This necessitates two kinds of rules.

First, pipeline systems have rules for minimum quality requirements (pipeline specifications) of the product to be received into the pipeline. Such rules serve to avoid contaminations that would deteriorate the quality of the entire blend, and to avoid substances that would cause problems in the pipeline itself. Such rules however do not provide for all products in the pipeline to be of the same quality, which would generally be unfeasible given that the product originates from fields with different hydrocarbon compositions in the reservoirs.

Second, many pipeline systems have rules for value adjustment for quality, in order to equalize the value of the product recovered with the product supplied by each transporter. For dry natural gas this can be based simply on the energy content of the gas, measured in joules or British thermal units, since natural gas generally obtains its sales value on the basis of energy content.

For crude oil pipelines, quality adjustment is more complicated, since different crude oils have differences in market value not coinciding with differences in energy content. Generally, crude oil buyers put a higher value per barrel on crudes which yield high fractions of the most valuable refined products, such as gasoil (diesel) jet fuel and petrol, in primary distillation. Residual fuel oil, on the other hand, is a much less valuable product. This is resolved by rather elaborate calculations of the relative value of crude oils moved in the pipeline, based on their calculated values in refining. The calculation may change from one period to another due to the changing market prices of the refined petroleum products. The outcome of the process is a set of adjustments according to which the suppliers of lower valued oil into the pipeline must compensate the suppliers of higher valued oil, bearing in mind that they will all get the same blend of crude oil at the end of the pipeline. The adjustment can be required as cash payment or in kind, the latter entailing that the supplier of higher value crude will receive more barrels than he supplied, and vice versa.

6.6 Marketing oil and gas

Some knowledge of how the produced oil and gas will be marketed is useful when considering possible offtake solutions. The ways in which these markets normally operate are quite different for crude oil and natural gas.

6.6.1 Crude oil: marketed quality

Crude oil is not a finished consumer product. With very few exceptions, crude oil is used in oil refineries, which convert crude oil to the petroleum products which consumers and businesses can use: petrol, jet fuel, diesel, heating oils, etc. The quality of crude oil differs between reservoirs of origin. The refining industry is configured to receive and handle such differences in crude oil quality, though individual refineries may be more or less suited to particular qualities of crude.

The crude oil that is transported from a producing field to refineries needs to be suited for transportation and to provide a reasonable degree of consistency in what refiners receive. The field processing requirements for this will be reviewed in the next chapter.

If the crude oil is to be transported in a pipeline system serving many fields, the oil entering the pipeline must conform to quality specifications set by the pipeline owner, in order to avoid problems with pipeline flow and to avoid contaminating the other oil in the pipeline with undesirable components.

Quality specifications are also agreed in crude oil sales contracts, but mainly buyers will rely on *crude oil assays* and previous refining experience to assess the oil's suitability for refining. A crude oil assay is a comprehensive analysis of a crude oil with respect to its refining properties. Crude oil assays for important oil fields and terminals are updated from time to time, since the crude composition can change gradually due to oil originating from different reservoirs being commingled in the pipeline and terminal systems.

Sales and pipeline quality specifications for crude oil normally include limitations on the following:

true vapor pressure (TVP);
Reid vapor pressure (RVP);
base sediment and water (BS&W);
salinity;
temperature;
hydrogen sulfide (H_2S) content.

Quantities of oil in international trade are usually measured in barrels. One barrel is 159 L.

6.6.2 Crude oil: sales contracting

For crude oil that is transported on ships, sales contracts tend to be of relatively short duration. Spot sales of individual cargoes are common, especially in the Atlantic basin. Term contracts with duration of 1–5 years are also common.

It is usually not necessary to secure oil sales for a field development before embarking on a field development, provided that it will be possible to export the oil on ocean-going ships and that the oil does not have a particularly difficult quality for refining purposes. For pipeline-bound crude or unusual oil grades, there may be a bigger need for long-term agreements as a condition for development, because the number of potential buying refineries will be more limited and the mutual dependence between oil producers and refiners stronger.

It is normally possible for the crude oil marketing function of an oil company to adjust the volume of sales to whatever quantities will be produced; hence, it should be possible to plan production based on optimal field management, rather than sales commitments. Unplanned variations in production may however cause problems and costs for the marketing function.

Crude oil sales contracts contain nomination and scheduling procedures, to determine the timing of each cargo loading. Such nomination procedures in sales contracts must be consistent with similar procedures applicable at the loading terminal.

The price in crude oil term contracts is usually agreed as a formula, where the price for each cargo will be calculated based on published market price assessments from agencies such as Platts (www.platts.com), around the time of cargo loading.

Contracting of shipping capacity is referred to as *chartering*. It may be done by either sellers or buyers of crude oil, depending on whether sales are made free on board (FOB), in which the buyer arranges for shipping, or Cost and Freight (CFR), in which case the seller arranges for shipping. Crude oil tankers can be contracted on long or short contracts (time charters and other forms), and there is also a considerable market in spot chartering of such ships for individual voyages. Certain operations requiring specialized ships, such as offshore loading ships, are usually chartered on long contracts.

6.6.3 Natural gas: marketed quality

The product known as *natural gas* in global trade and among consumers is gas that has been processed to meet the quality specifications that apply to downstream sales as well as to the pipeline systems that carry gas from different origins. Quality specifications for sales gas normally include limitations on the following:

water content;
water dew point;
hydrocarbon dew point;
hydrocarbon composition;
heating value (energy per gas volume specification);
contaminants (H_2S, CO_2, N_2, COS, mercaptans, mercury, etc.).

Unlike crude oil, natural gas is a finished consumer product. Once processed to sales specifications (which can be accomplished at the producing field or in gas processing plants), gas requires no further substantial processing before being consumed as a fuel by businesses, households, or power plants.

In chemical terms, natural gas is a rather simple product, consisting of approximately 90% of methane (CH_4), the smallest hydrocarbon molecule. The remaining part is mainly ethane, propane, CO_2, and sometimes nitrogen. Natural gas obtains its value from its energy content, and sales contracts usually contain provisions to the effect that the price to be paid is a function of the amount of energy provided.

For commercial purposes, quantities of natural gas are identified by energy content or by volume. Energy is measured in joules, British thermal units, or their multiples. One megajoule (MJ) is 1 million joules. One million British thermal units is usually written as 1 mmBTU, and is a common pricing unit for gas internationally. The energy measurements referred in international trade usually refer to the gross calorific value (GCV; also known as the higher heating value) of the gas, which for a given quantity of gas is about 10% higher than the net calorific value. The difference is related to condensation of water steam generated in the combustion process.

The volume measurements of natural gas are standard cubic feet (scf), standard cubic meter (Sm^3), and normal cubic meter (Nm^3). The difference between a standard cubic meter and a normal cubic meter relates to the temperature at which the gas is measured as a cubic meter: 15 °C for Sm^3; 0 °C for Nm^3. Following are approximate conversion factors:

1 Sm^3 = 35.3 scf;
1 Nm^3 = 37.2 scf;
1 $Sm^3 \approx$ 40 MJ GCV (depending on gas quality);
1 BTU = 1055 J;
1 mmBTU = 1055 MJ.

6.6.4 Natural gas: sales contracting

Unlike crude oil, natural gas from producing fields is commonly sold under very long-term contracts, often longer than 20 years. Furthermore, such sales need to be secured before a final commitment is made to develop a field with large gas reserves. The reason for this long-term orientation is the strong and costly reliance on infrastructure linking buyers and sellers, usually by pipelines, or alternatively as LNG transport chains (Figure 6.8). The flexibility of being able to reach a large number of potential buyers for ship cargoes, which is the basis for the short-term orientation of crude oil trading, is largely lacking in natural gas, although developments in LNG trade since the 1990s may point to more short-term orientation in the future.

Pipeline gas contracts tend to provide for long-lasting and essentially stable rates of delivery. Contracts provide for an annual contract quantity (ACQ) and a daily contract quantity (DCQ), which the buyer will nominate within specified limits. Gas contract prices tend to be

Figure 6.8: The LNG tanker "Arctic Discoverer" at the Cove Point terminal, USA. Photo: Statoil/R. Lindefjeld.

agreed as a formula linked to reported market prices of other forms of energy, notably crude oil or refined oil products such as heavy fuel oil and gasoil.

Demand for natural gas is highly seasonal in the major consuming nations in the Northern hemisphere, since a large portion of the gas is used to heat buildings. Spot prices for natural gas have a strong tendency of being higher in winter than in summer. A big issue in gas contract negotiations therefore tends to be the degree to which the buyer will be allowed the flexibility to take larger gas volumes in winter than in summer, an issue which also has important implications for the production and transport systems upstream. Power generation and industrial process applications are also major gas-consuming sectors.

6.6.5 Marketing of natural gas liquids

Prior to being sold, the NGLs are usually fractionated into the individual products at a gas processing plant. Following are the products which may be obtained from the NGL component, and their main applications:

Ethane (C_2H_6): Used as a feedstock for petrochemical cracking to make ethylene. Ethane is in many cases not produced as a separate product, but left as a component of the sales gas.
Propane (C_3H_8): The largest NGL product. Propane has many applications in heating, automotive fuel, and as a feedstock for petrochemical cracking.
Butane (C_4H_{10}): Butane has similar applications as propane, and is also used as a blendstock in petrol. Propane and butane are jointly known as LPG. Some gas processing plants split butane into normal butane and *iso*-butane, which have somewhat different industrial applications.
Light condensate (C5+): Light condensate is a liquid with a final boiling point of not more than 180 °C. It is obtained mostly by processing associated gas from oilfields. Resembling light refinery naphtha, it is used mainly as a petrochemical cracker feedstock and in certain other processing applications.
Heavy condensate: Heavy condensates are obtained from processing non-associated gas. It is a liquid with a yellowish appearance and a final boiling point usually between 180 °C and 360 °C. Some petrochemical crackers can use heavy condensates as feedstock; otherwise, they are sold to refineries and processed much as a light crude oil.

There are well-established international markets for propane and butane, including trade in ship cargoes (LPG gas carriers). These markets operate similarly to oil trade in that most sales are done on relatively short-term contracts and some spot trade. Light condensates are traded as part of the market for naphtha, which has similar characteristics.

Ethane as an isolated product is a much smaller and specialized market. Any petrochemical cracker using ethane would need special facilities to handle it (which most European crackers do not have, though it is more common in the USA), and therefore an ethane project as part of gas processing would require long-term agreement between sellers and buyers. Heavy condensate is also a specialized market; it can find buyers in refineries, but more valuable applications in petrochemicals may require careful preparations.

6.7 Summary

Alternative solutions for hydrocarbon offtake may include pipeline transport, ship transport, and local gas conversion. Their relevance depends on the product specification, the capacity of the processing facilities, the transport distance, and volumes.

Different types of onshore terminals are installed for a number of different purposes related to the receipt, processing, and onward transfer of oil, gas, and NGLs.

Crude oil is used by oil refineries, which convert it into petroleum products. Crude oil for sale as ship cargoes is sold on short- to medium-term contracts as well as spot sales, which generally do not have to be arranged before a development project is initiated. Crude oil sold into pipelines may require more long-term marketing. Natural gas is mostly sold under very long-term contracts, which generally need to be in place before a development relying on such sales is committed. NGLs are valuable by-products of gas processing that can usually be sold on short- or medium-term contracts as ship cargoes.

Overall field design and support facilities

Previous chapters have described many kinds of equipment which are necessary for oil and gas production: drilling rigs, wells, pipelines, and processing equipment. These are the installations which come in direct contact with the oil and gas being produced. Oilfield operations also require a number of other installations of a more auxiliary nature, without which the core processes cannot function. They are often referred to as "facilities." There is also the challenge of how to locate the installations, particularly in offshore environments.

The chapter first introduces facility requirements which may apply irrespectively of where production is located. Requirements for onshore developments are then briefly reviewed, before addressing more extensively the particular challenges of offshore oil and gas production.

7.1 General requirements for field design and support facilities

This section addresses types of installations required for most oil and gas production projects both onshore and offshore.

The overall plan for development and operations of a field must lay out the physical configuration of field installations, to ensure that all required functions are covered in a rational and consistent manner.

7.1.1 Wells and their connections to process facilities

The functions of well construction and flow assurance are covered in previous chapters, and as reviewed there, they require installations such as wells and flow lines.

Locations for wells and wellhead installations need to be specified. In traditional onshore field developments, wells were often drilled vertically, requiring a separate wellhead location for each well with surface tie-back to a processing facility or gathering point. This is hardly an option for offshore developments, which usually require many wells to be drilled directionally from one platform or floating facility (Chapter 3). Increasingly this is a preferred solution also for onshore developments, which reduces the need for infield lines and environmental disturbances on the surface.

Wellheads must be configured so as to allow later access with relevant equipment for well testing, well workover, and other anticipated operations.

Infield manifolds must be installed to collect wellhead fluids and distribute injection fluids. Infield pipelines are required for wellstreams, injection fluids and test lines, as well as control cables. Wellhead lines may have to be heated, insulated, or injected with inhibitors to avoid wax and hydrate problems (Chapter 4).

7.1.2 Injection systems

Producing fields may require facilities for injecting gas or water into reservoirs for pressure management or disposal, and for injecting chemicals and components into the production system for the purpose of flow assurance or production enhancement.

Gas injection into underground formations is done with dedicated injector wells. It may be done for the purpose of reservoir management for optimized oil recovery, or in order to dispose of associated gas that has no viable outlet to market when produced (Chapter 2).

Water injection into underground formations is also done with dedicated injector wells, and for the two possible purposes of optimal reservoir management and disposal of produced water. Such produced water may otherwise require extensive treatment before release to the surface environment, to avoid pollution. Water injected for reservoir management purposes may also be obtained from the ocean or other environment. Water may require treatment before injection in order not to cause problems in the reservoir or facilities.

CO_2 injection into underground formations is also done with dedicated injector wells, and for the two possible purposes of optimal reservoir management or disposal of a substance whose release to the atmosphere would be detrimental to the environment. CO_2 injection has long been practiced for increasing oil production from certain fields in TX, USA, using CO_2 that is extracted for this purpose from separate gas reservoirs having a high CO_2 content. The Weyburn oilfield in Canada optimizes oil production using CO_2 that is obtained from a coal gasification operation across the US border. The Norwegian gas/condensate fields Sleipner and Snøhvit have 5–10% CO_2 contents in their wellstreams; this CO_2 is extracted and injected into the underground for the sole purpose of avoiding its release to the atmosphere. Such release to the atmosphere is discouraged by a hefty Norwegian tax on CO_2 emissions. A similar CO_2 injection scheme operates at the In Salah gas field in Algeria, and is planned world wide for several other large gas fields having large reservoir contents of CO_2. The CO_2 must be separated from the natural gas, for which amine absorption is so far the most common technique (Chapter 5).

Injections into reservoirs for improved oil recovery can include other substances than gas, water, or CO_2. The three main techniques are *thermal techniques*, *chemical techniques*, and *miscible processes*. Thermal techniques aim to reduce the viscosity of heavy crude, thereby improving the crude's mobility. This can be done by injecting steam into the reservoir, or more ambitiously by igniting a mixture of hydrocarbon gases and oxygen (*in situ* combustion). Chemical techniques change the properties of either the displacing fluid (usually water) or the oil. Examples of chemical techniques are polymer flooding and surfactant flooding. Polymer flooding aims to increase the viscosity of the injected water, and surfactant flooding aims to reduce the amount of oil left behind in the pore spaces of the reservoir rock. Miscible processes are aimed at recovering oil which would normally be left behind as residual oil, by using a displacing fluid that mixes with the oil (Jahn et al., 1998). Injection procedures for optimized oil recovery are further described in Chapter 2.

Injection of chemicals to production wells is performed during well workover operations to reduce the constraints which the well completion may impose on the productivity, such as scale formation, emulsions, asphaltenes, and wax. Hydrochloric acid (acid wash) is one type of chemical used for this purpose.

Injection of chemicals into production pipelines may include such chemicals as monoethylene glycol (MEG), methanol, scale inhibitors, and emulsion breakers. Chemical injection into pipelines is performed for flow assurance purposes (Chapter 4), such as to dissolve asphaltenes, wax, emulsion, hydrate, and scale, and for corrosion control.

7.1.3 Process, utilities, and metering facilities

The need for processing the wellstream into products that can be exported from the production site was reviewed in Chapter 5.

Depending on the offtake solution, there may be a need for storage of oil, possibly also for gas, at the field. Pipeline offtake usually eliminates the need for storage in the field. For oil pipeline systems, the required flexibility can be provided by downstream storage tanks at refineries or oil terminals. In large gas pipeline systems, line packs provide some flexibility for handling short-term fluctuations in production or consumption. Line pack refers to the possibility of varying the amount of gas in the pipeline by varying its pressure. This is possible for gas, but not for oil in a pipeline.

Oil and gas production operations need substantial amounts of power, which for most purposes needs to be supplied as electricity. For certain power-intensive applications such as drilling, gas compression (for injection or pipeline export), oil pumping, and water pumping, power can be supplied either as electricity or directly from a combustion engine. Diesel engines have a number of applications in oil and gas fields, including as power generators. If there is gas production, it is often attractive to use some of it for generating power for the field's requirements, especially if it is difficult to find attractive offtake for associated gas. Power generation from gas can be done with a reciprocating engine, a gas turbine, or in a steam boiler with a steam turbine. Gas turbines are compact and can be quite flexible in the fuel quality they use, but tend to be less fuel efficient than reciprocating engines.

It may be relevant in some cases to connect an oil and gas field to the general electricity grid, especially onshore, but it has also been done for some offshore fields via subsea cables. Such connection would normally be made for the purpose of obtaining grid electricity as the main power supply for the field, but it can also be relevant for the field to supply surplus electricity to the grid or to local users.

Measurement packages will be required for metering produced fluids. Measurements are made for two purposes: production management (including reservoir management) and fiscal metering as a basis for transactions between parties.

7.1.4 Logistics, quarters, and safety facilities

Onshore fields require access by road and/or other relevant transport, and in many cases an air strip is useful. Bringing in heavy equipment for drilling, processing, etc., as well as large amounts of pipe and other supplies must be planned with due consideration to available roads and other infrastructure in the region. Intrafield access to well sites, etc., must also be planned.

Offshore field installations are generally constrained by space and weight limitations for storing equipment on board. A shore supply base will usually be required, using supply vessels between the base and offshore installation, and with access for freighter ships and land transport to bring supplies from manufacturing sources to the base. The offshore installation must then include cranes and adequate space for lifting supplies from supply vessels to the installation, and for returning items to shore.

Offshore installations generally also include a landing pad for helicopters, for bringing personnel and some supplies on and off the installation. An adequate onshore heliport facility is also required.

Accommodation must be provided for persons who need to stay at the field, including company employees as well as contracted personnel. Standards in this regard vary greatly in the industry, with the Norwegian offshore sector being noted for providing high standards of accommodation.

One should bear in mind that oil production deals with hazardous substances and lots of heavy moving equipment; therefore, the field installations require a number of systems for avoiding disasters and protecting lives. They include gas and fire detectors, emergency shutdown procedures, fire extinguishing systems, fire walls, first aid and medical services, life boats (offshore), etc. As a consideration for field installation design, there should be segregation by distance and by physical barriers between the most hazardous parts of installations (usually processing equipment) and the parts likely to contain most people (usually accommodation quarters). Chapter 10 will review safety management principles.

7.2 Onshore developments

Figure 7.1 illustrates a generalized outline of an onshore oil and gas development. Wells are connected via manifolds and infield flow lines to a central processing facility, from which one trunk line for oil and one for gas provide the offtake. Figure 7.1 indicates a traditional configuration of

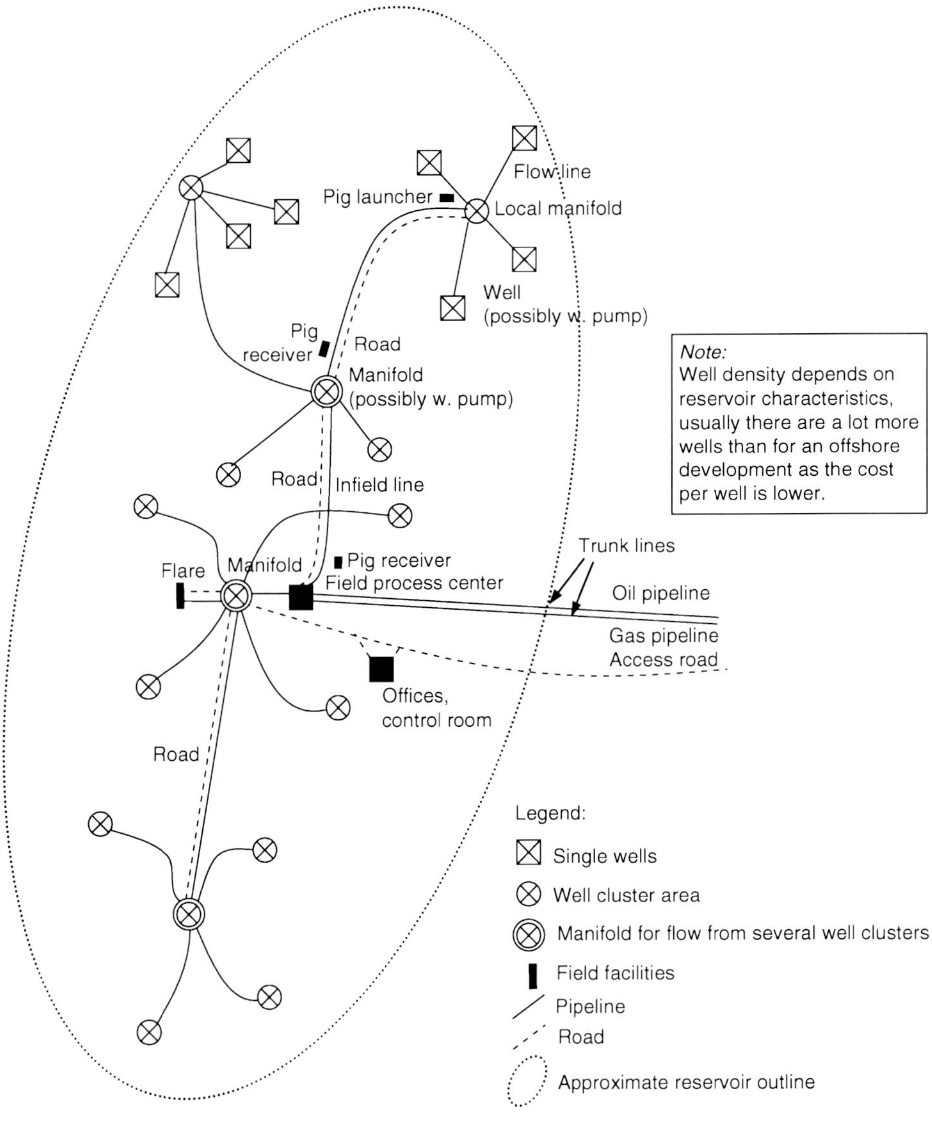

Figure 7.1: Generalized onshore development layout.

vertical wells from individual wellhead locations. Figure 7.1 does not indicate any gas or water injection, which would have required additional wells and flow lines.

Onshore developments require considerable amounts of civil construction works. This has environmental impacts, but can also benefit local communities in the form of improved roads, etc. By consulting with relevant authorities and local community representatives, the field developer (operator) may achieve a project that is compatible with the needs of residents in the area. Minimizing releases to air and water is a general environmental requirement.

Lower drilling costs often lead to a higher number of wells in an onshore development than would be the case in a similar offshore reservoir. The reservoir and drilling engineers will determine the drainage pattern and optimize the number of wells. Considerations in this regard include the following:

- reduce the impact on the environment;
- reduce the drilling and completion costs;
- consider possible multiple completion of wells;
- optimize reservoir management, including re-evaluating the need for injection;
- conduct reservoir simulation studies;
- maintain the wells and re-use parts of wells which are not in production.

A general trend in onshore development is to reduce the number of locations from where the wells are drilled, through use of deviation drilling of wells (not reflected in Figure 7.1). This trend is in general considered to be more environmentally sound than drilling from a larger number of locations.

7.3 Offshore developments

Developing offshore fields requires different development strategies compared to an onshore environment. The main difference in drainage strategy relates to optimizing the well pattern using a lower number of wells and wellhead locations than would normally be used at a corresponding onshore field. Whereas wells can be arranged in a simple grid pattern onshore, they must be drilled from a small number of locations offshore. Such locations can be platforms or subsea templates. The determination of drainage points is therefore a more critical issue offshore than onshore.

The cost of platform support structures increases considerably with increasing topside weight. Avoiding any unnecessary process equipment offshore is therefore an important consideration. Any increase in processing equipment offshore further adds to overall life cycle costs through added needs for instrumentation, equipment maintenance, and manning, which are more expensive offshore than onshore.

Based on the required well drainage pattern and the export product specifications, a field layout including main installations must be determined. Offshore development concepts include the following:

- fixed platform;
- floating platform;
- subsea wells to a fixed platform;
- subsea wells to a floating platform;
- subsea wells to a ship unit;
- full wellstream from subsea templates in pipelines to land.

In all but the last listed concept, oil offtake can be by offshore loading of oil tankers or by pipeline to shore.

A fixed platform may be one integrated platform providing all necessary functions. An alternative fixed platform concept is to have adjacent separate platforms providing different functions and linked by gang bridges, often three platforms for wellheads, processing, and quarters. A traditional field development for shallow waters is shown in Figure 7.2.

Figure 7.3 shows the most modern type of offshore development solutions, implemented for the development of the Ormen Lange field offshore Norway. It features the transfer of the full wellstream from subsea installations to a shore facility, requiring no offshore installation above sea. The Snøhvit gas/condensate development in northern Norway is also one such development.

Alternatively, the solution would be a stand-alone deepwater development with processing on a floating platform, oil offtake by ship, and gas offtake by pipeline to an onshore gas processing plant. The Åsgard field in Norway is such a development.

Platforms are generally classified by their mechanical construction. The four main types are (Jahn et al., 1998):

- steel jacket platforms;
- gravity-based platforms;
- floating platforms (including floating rigs);
- minimum facility systems.

Included in minimum facility systems are subsea installations and production ship solutions. Artificial islands may be regarded as platforms, combining features of onshore and

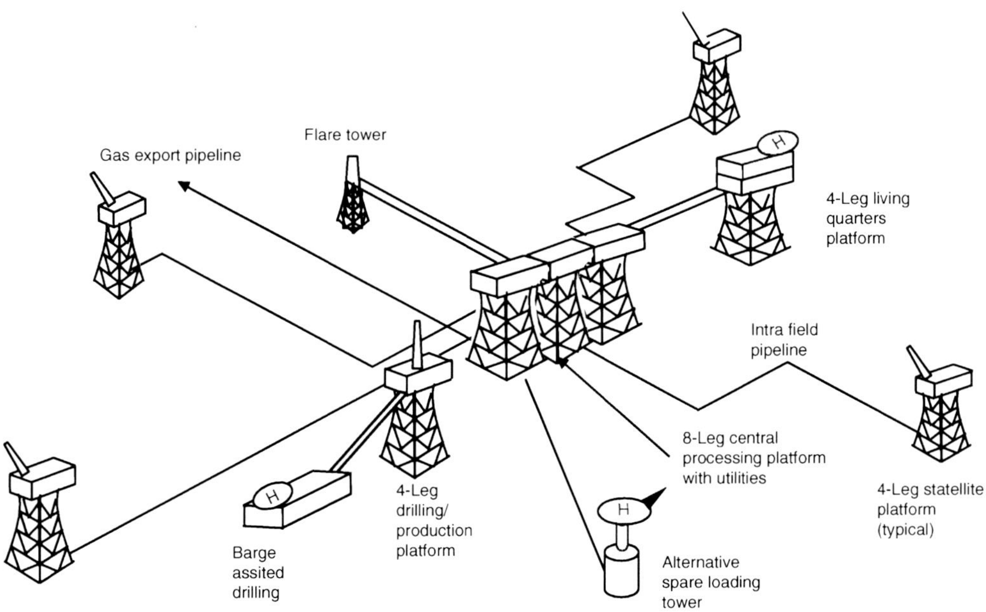

Figure 7.2: Traditional field development solution for shallow waters (The central processing platform could have 4, 6 or 8 legs with associated pile groups).

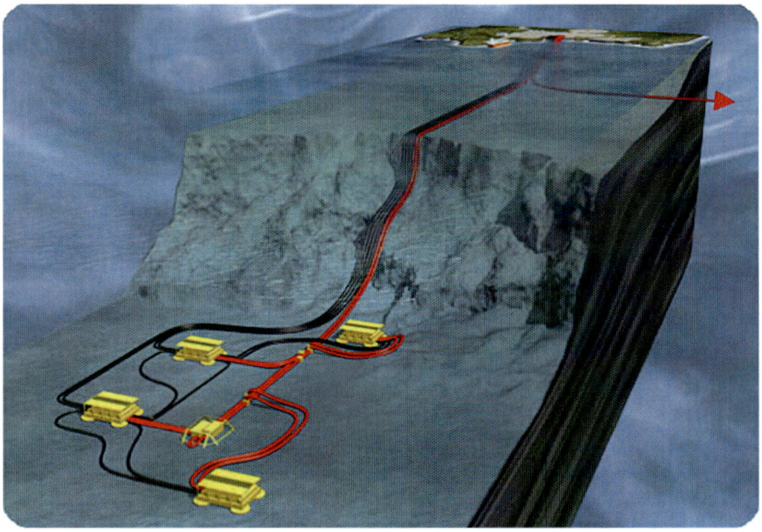

Figure 7.3: Subsea development of the Ormen Lange field in deep water with wellstream to shore (illustration: Statoil).

offshore facilities. Some of the fixed platforms, floating platforms, and production ship solutions available are illustrated in Figures 7.4 and 7.5.

Steel jacket platforms are the most common platform type, and are applicable for a large range of sea conditions. These include both the calm West African seas and the rough North Sea. Steel jackets are used to depths of about 150 m, and may support structures that are 30 m above the sea level. Steel jackets are built from welded steel pipes. They are built onshore, and floated out to sea. Once in place, they are pinned to the seafloor with steel pipes (piles). Steel jackets can weigh up to 20,000 tons, and support similar weights (Jahn et al., 1998).

Concrete and steel gravity platforms may be used in similar water depths as steel jackets. The main difference between steel jackets and gravity structures is that the gravity structures rely on their weight to secure them on the seabed. This eliminates the need for piling in hard seabeds. Concrete structures are built with hollow ballast tanks and hollow concrete legs. This allows the platform support structures to be floated into position without the use of a barge, and be sunk in place on location. The caissons of these platforms may be used as temporary storage for crude oil, and some platforms may accommodate storage in excess of 1 million barrels (Jahn et al., 1998).

Floating platforms are mainly used in deep waters where fixed platforms would be vulnerable or very expensive. Types of floating platforms are tension leg platforms (TLP), spar platforms, deep draft floaters, and semi-submersibles, illustrated in Figures 7.9–7.12.

The main platform support solutions have the following characteristics which may influence their selection for field development projects:

1. *Fixed steel platforms* (Figure 7.6):
   - well-established technology;
   - for limited water depth;
   - platform wells;
   - rigid (supported) risers;

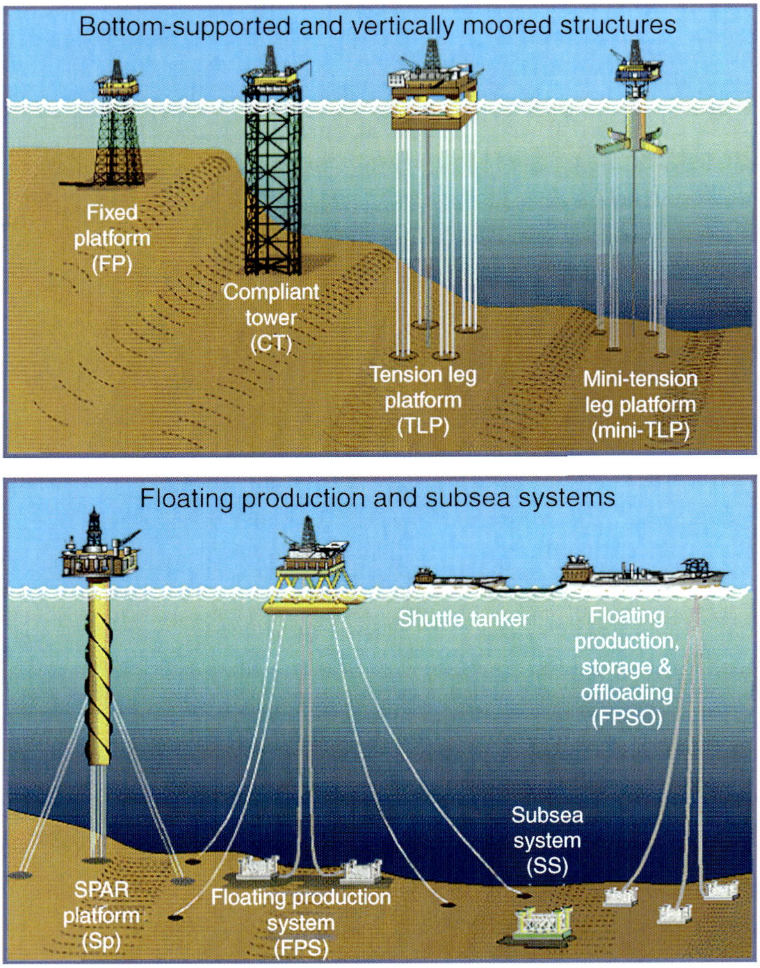

Figure 7.4: Some fixed platform, floating platform, and production ship solutions (from the homepage of Minerals Management Service, basic options used in constructing the Gulf's deepwater permanent production platforms. Illustration: Minerals Management Service, US Department of the Interior.

- comprehensive marine operations required;
- offshore installation of topsides required;
- no oil storage.

2. *Jack-up platforms* (Figure 7.7):
   - well-established technology;
   - ideal for shallow water;
   - limited water depth;
   - platform wells;
   - rigid tensioned risers;

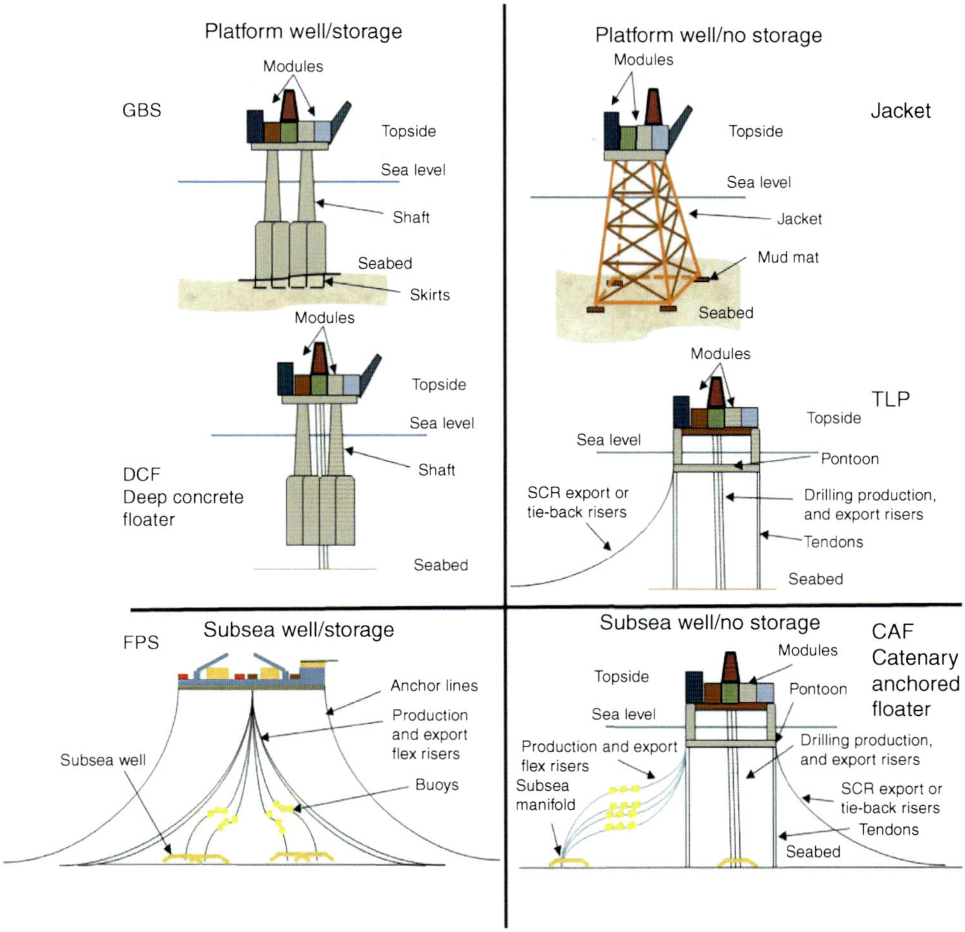

Figure 7.5: Some platform concepts and their well solutions. GBS: gravity base structure; TLP: tension leg platform; FPS: floating production and storage; SCR: steel catenary riser.

- simple marine operations and self-installing platform;
- at-shore installation of topsides;
- possibility for subsea oil storage tank;
- vulnerable to the seabed conditions;
- movable for re-use at another location;
- used extensively world wide for drilling wells.

3. *Fixed concrete platforms* (Figure 7.8):
   - well-established technology;
   - robust long-life structures;
   - limited water depth;
   - platform wells;
   - rigid (protected) risers;
   - deepwater construction site and transportation route may be required;

Figure 7.6: The Brage steel jacket platform in the North Sea. Photo: Statoil/Kjetil Alsvik.

- comprehensive marine operations required;
- in-shore installation of topsides;
- oil storage possible.

4. *TLP* (Figure 7.9):
   - well-known technology, but need for fine-tuning hull and mooring configuration;
   - complex dynamic behavior;
   - suitable for deep water, but in ultra-deep water the tethers will become too heavy;
   - platform wells;

Figure 7.7: The jack-up platform Mærsk Giant at the Yme field in the North Sea. Photo: Statoil/ Øyvind Hagen.

Figure 7.8: Fixed concrete platform during tow to field: Troll A platform in the northern North Sea. Photo: Statoil/Leif Berge.

Figure 7.9: The Snorre TLP platform. Illustration: Norsk Hydro.

- top-tensioned (exposed) rigid risers;
- comprehensive marine operations required;
- at-shore installation of topsides;
- no oil storage.

5. *Spar platforms* (Figure 7.10):
   - proven technology;
   - limited deck area;
   - conventional mooring;
   - suitable for deep water;
   - complex dynamic behavior, particularly in case of large currents;

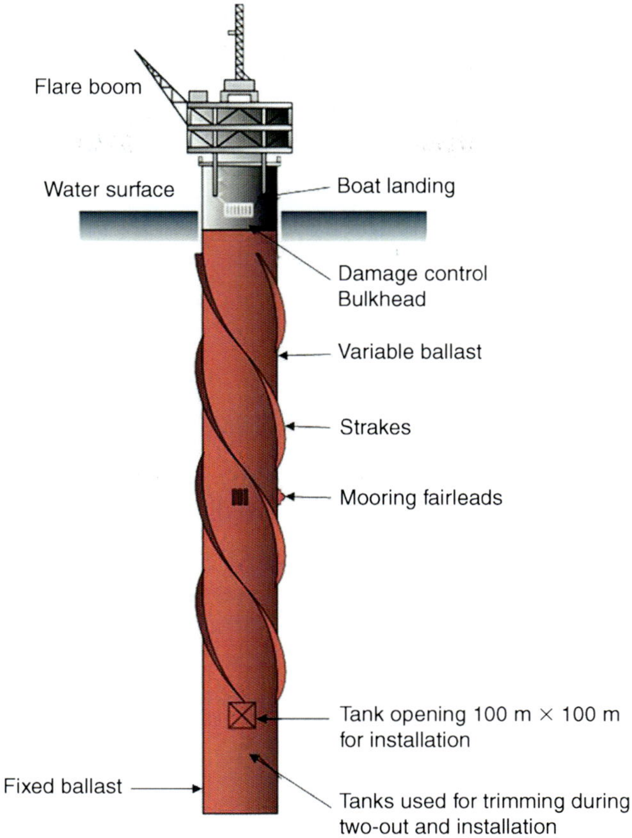

Figure 7.10: A spar platform for deep water.

- platform wells;
- top-tensioned (exposed) rigid risers;
- risers protected in wave zone;
- complex marine operations required;
- offshore installation of topsides required;
- possibility for some oil storage, but not tried.

6. *Floating platform (semi-submersible)* (Figure 7.11):
   - proven technology;
   - possibility for a drilling rig;
   - conventional mooring;
   - suitable for deep water;
   - well-understood dynamic behavior;
   - subsea wells;
   - flexible risers;
   - simple marine operations;
   - at-shore installation of topsides;
   - no oil storage.

Figure 7.11: The Kristin semi-submersible. Illustration: Statoil/Even Edland.

7. *Deep buoy-type platform*:
   - unproven technology, but similar to semi-submersibles;
   - possibility for a drilling rig;
   - limited deck area;
   - conventional mooring;
   - suitable for deep water;
   - well-understood dynamic behavior and large motions;
   - subsea wells;
   - flexible risers;
   - risers protected in wave zone;
   - simple marine operations;
   - at-shore installation of topsides;
   - oil storage possible.
8. *Ship-shaped units* (*floating production, storage and offloading* (*FPSO*), *floating production, storage, drilling and offloading* (*FPSDO*), *and floating storage unit* (*FSU*)) (Figure 7.12):
   - variants include FPSO, FPSDO, and FSU;
   - proven technology;
   - large deck area;
   - conventional mooring;
   - suitable for deep water;
   - well-understood dynamic behavior and large motions;
   - weathervaning;
   - subsea wells;
   - flexible risers and swivel;

Figure 7.12: The Norne FPSO. Photo: Statoil.

- congested riser area;
- simple marine operations;
- at-shore installation of topsides;
- oil storage usually included.

The water depth and the number of wells required are key determinants of the type of platform concept (if any) which can be suitable for a particular offshore development.

There has for many years been a tendency for production activities to move into deeper waters, as many of the easily accessible oil and gas fields in the world are drying up. This has triggered the development of new concepts such as subsea to shore developments, in which the full wellstream is transported to shore, as well as ship-shaped production units.

7.4 Subsea facilities

Subsea facilities are installations placed entirely under sea. The capabilities and applications of subsea facilities have increased over a number of yeas. Subsea installations are often used to produce from locations where a separate platform would be unnecessary and uneconomic, and which cannot easily be reached by directional drilling from an existing platform.

Mature oil and gas provinces are areas in which oil and gas have been producing for a number of years, existing production tends to decline, and where it is often assumed that most of the *major* hydrocarbon accumulations have already been discovered. Remaining accumulations are often too small to justify stand-alone offshore platform developments. Subsea production systems may offer an attractive alternative development option. By tying wells located at the seabed to existing infrastructure such as platforms and pipelines, hydrocarbon accumulations that would be uneconomic as stand-alone developments can be profitably exploited.

Also, subsea wells may be tied back to production ships or even all the way to onshore processing facilities and terminals. Operators of fields that are beyond peak production may have excess capacities available at their platforms and offtake systems. With such excess capacity, they may find it attractive to host production from nearby fields, for which they can be compensated with tariffs on the third-party hydrocarbons they handle (Amin et al., 2005).

7.4.1 Main components of subsea systems

A subsea system usually has the following main parts:

- *The subsea well system*: Consists of the downhole completion system, the subsea tree system, and the rig/vessel system.
- *The subsea production system*: Consists of the subsea control system, the support and protection structures and the manifolds, intervention systems, and subsea processing systems.
- *The subsea pipeline system*: Consists of the subsea flow lines, the tie-in systems (including the umbilicals, which may include, e.g., control cables as well as lines for injection of fluids at the wellhead), the riser system, and the subsea pipelines.

The subsea system components are shown in Figure 7.13.

7.4.2 Wet trees, manifolds, and templates

The most basic form of subsea installation is a wellhead with a subsea tree, connected to a production facility by pipelines and umbilicals. A subsea tree (also called a *wet tree*) is a series of valves and gauges located at the top of the well. A control module situated at the subsea tree

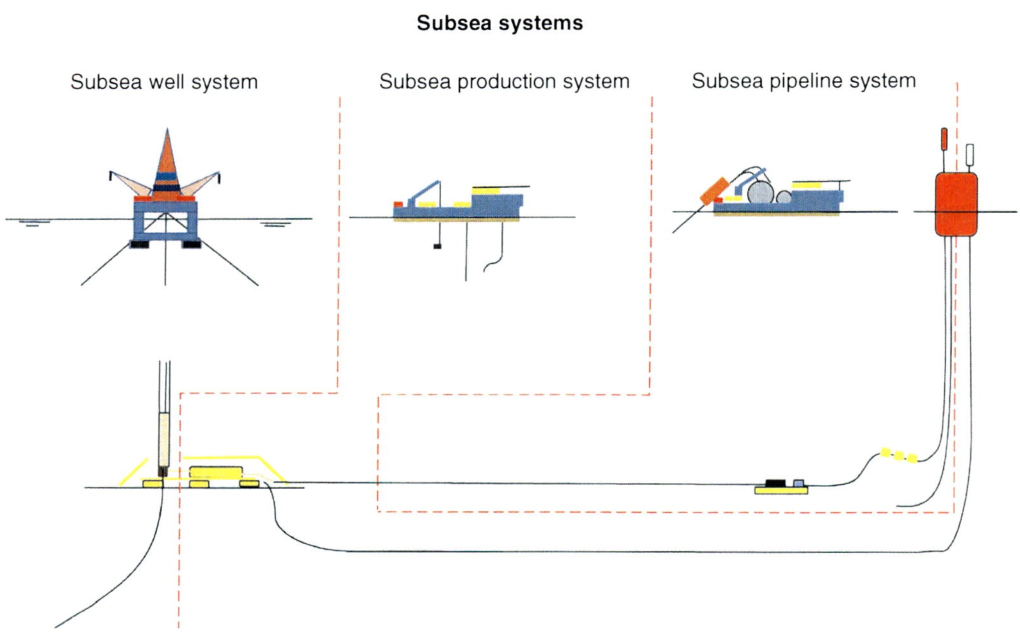

Figure 7.13: Subsea system components.

allows the facility to be remotely operated. Water injection, chemical injection, and gas lift can be provided from the host facility.

When several subsea wells are tied together, a *manifold* is used as a common focal point that collects, commingles, and exports the flow to a surface-gathering facility. Flow lines and umbilicals connect each well to the manifold. From the manifold back to the surface host facility, a set of pipelines, umbilicals, and risers is required. Subsea manifolds are generally flexible and cost effective. They are usually made from tubular steel parts, and include such components as pipes, chokes, valves, and flow meters. It is common for subsea trees to be mounted directly at the manifolds, and not on the individual subsea wells (Jahn et al., 1998).

Subsea *templates* (Figures 7.14 and 7.15) have the same function as manifolds, but are different with regard to the location of wells. On templates, the wells themselves are located directly below the structural frame, which protects and connects them. Hence, subsea templates are more than mere focal points; rather, they are structures that allow several wells to be drilled and completed from one location.

7.4.3 Umbilicals, control systems, and surveillance

As subsea systems are remote from their host production location, which may be located on a platform or onshore, they require remote control and surveillance. Modern subsea components are controlled via *umbilicals*, which may consist of the following types of cables and lines:

- *Electric cables*: Used for powering the control system and some actuating components, and for communication. Signals are sent back to the surface to provide information on the status of valves and chokes, flow rate measurements, pressure measurements, and temperature measurements.
- *Hydraulic cables*: For operating actuating components such as valves, etc.

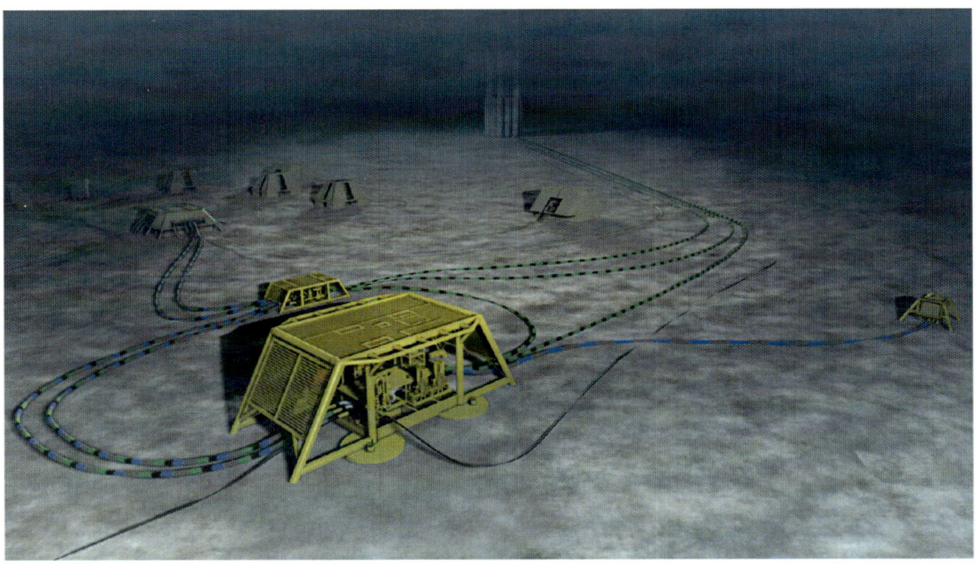

Figure 7.14: Subsea templates for a North Sea field. Illustration: FMC Kongsberg/Statoil.

Figure 7.15: Preparations for installing a well template. Photo: Statoil/G.O. Johansen.

- *Chemical lines*: For transporting substances that are to be injected into the production pipeline, such as Monoethylene glycol (MEG), methanol, scale inhibitors, and emulsion breakers.
- *Optical cables*: A modern alternative to electric cables for communications with the subsea equipment.

The umbilicals connect the subsea *control distribution unit* (CDU) and the surface (or onshore) *control station*. The umbilicals transfer information about the status of subsea equipment such as temperature measurements, pressure measurements, choke settings, flow rates, etc., back to the surface. Based on this information, the surface control station will return signals that will initiate appropriate corrective measures if the status of the subsea measurements is unsatisfactory.

7.4.4 Subsea processing

Subsea processing is a set of technologies of which some have been implemented, but which are still under development. They may provide the following functions:

- gas separation;
- liquid separation;
- gas compression;
- liquid pressure boosting;
- desanding;
- gas treatment;
- injection to reservoirs.

Subsea processing techniques are expected to yield a number of advantages, notably a significant potential for improved oil recovery (IOR) particularly in deep water. It can enable new development solutions, improve the utilization of existing infrastructure, and enable longer tie-backs from subsea to shore solutions. It can also avoid discharges to sea.

Several subsea processing solutions have been implemented or are planned for the Norwegian offshore, representing important steps forward in technology development. They include subsea separation (Tordis field), wellstream compression (Åsgard fields), and water injection (Tyrihans field).

7.4.5 Subsea intervention

Most wells require some intervention during their lifetime, in order to extend the life of the wells. Following are examples of well interventions (Christie et al., 1999):

- installation or service of subsea control valves;
- change of gas-lift valves;
- production logging;
- change of broken tubing;
- removal of scale and paraffin;
- perforation of new sections in the reservoir;
- cementation of existing perforations to shut off water flow;
- application of chemical treatments (e.g., acid).

Traditionally, interventions have been accomplished by the use of a drilling rig and a marine riser with diver assistance. However, returning a drilling rig to a subsea well to perform intervention tasks is a very expensive option. Therefore, the industry has investigated more cost-effective methods for subsea intervention.

Some light intervention tasks are performed by *remotely operated vehicles* (ROVs). Typical ROV intervention tasks would be visual inspection of subsea equipment, valve operations, installation operations (wire handling and tie-ins), non-destructive testing (crack testing), cleaning, maintenance, and repair.

Light well intervention (LWI) technology is a riser-less technology that involves lowering intervention equipment directly into a pressurized well by means of a wireline paid out from a dynamically positioned vessel, usually a ship. A key element is a lubricator system used to ensure a tight seal. Recognizing that wireline intervention may cut the cost by as much as 50% in comparison to conventional rig operations, several qualification campaigns have been carried out. As a result, riser-less well intervention has become part of the standard subsea tool kit. Further research and development are aimed at using composite cables and coiled tubing operations, and increasing the lubrication length to extend the capabilities of wireline tools.

7.4.6 Subsea technology in Arctic conditions

Subsea technology in Arctic conditions is presently used at the Terra Nova and White Rose fields on the Grand Banks off the eastern coast of Canada, in 120-m water depth. Oil production wells were predrilled from a semi-submersible drilling platform. Wellheads and production manifolds are placed in excavations in the seafloor, for protection from icebergs. A network of more than 40 km of flexible flow lines is used to transfer hydrocarbons to and from the wells. Produced gases are separated from the oil and re-injected into the reservoir for possible future extraction.

The connection between the FPSO vessel and the subsea flow lines is a "spider buoy" at the lower part of the turret. The spider buoy is the mooring point for the FPSO, and the pathway for fluids flowing between the FPSO and reservoir. The spider buoy has a quick-disconnect feature, allowing the FPSO to disconnect and leave the area in an emergency situation (to avoid colliding with an ice berg). Crude oil is offloaded from the FPSO onto large shuttle tankers, which move the oil to market.

At the Terra Nova location, drifting icebergs can occur without any pack ice cover. The field installations rely on the capability to disconnect the riser system should one fail to tow away a threatening iceberg. In contrast, the nearby Hibernia development relies on a massive concrete gravity structure for iceberg protection.

In contrast to ice bergs, pack ice is an extensive ice cover which moves with water currents. Such conditions require somewhat different solutions than those described above for icebergs in open water at Terra Nova (Figure 7.16). Drifting pack ice may require the use of propulsion to head against the ice drift direction, allowing the vessel to withstand the forces from drifting ice features. This may be supplemented with icebreaker assistance and the capability of disconnecting and temporarily leaving the site, should the ice situation be unmanageable.

Subsea developments in the Arctic may experience considerable downtime during drilling, which may be restricted by ice conditions during large parts of the year. Use of drill ships in combination with ice feature management may extend the drilling season. The need to get free from the ice before the seas freeze must be incorporated in plans. This is a particular necessity for jack-ups, which generally have low tolerances for impacts from ice. Operational limitations with considerable production downtime may be anticipated, especially if well maintenance is required.

Figure 7.16: Terra Nova field layout (courtesy of Petro-Canada).

Table 7.1: Suggested application of subsea development concepts for different Arctic conditions.

Water depth	Ice conditions			
	Mild ice conditions (e.g., Bohai Bay conditions or Cook Inlet conditions)	Drift ice features (e.g., Banks off Newfoundland or northeastern Barents Sea)	Intermediate ice conditions (e.g., Pechora Sea conditions, North Caspian Sea, or near-shore Beaufort Sea)	Heavy ice conditions (e.g., Kara Sea conditions)
Very shallow, <10 m	Artificial island	Artificial islands or use of caisson-type structures	Caisson type of ice-resistant structures or use of gravel island for the shallowest part (in general for <5 m depth)	Caisson type of ice-resistant structures
Shallow, i.e., less than depth of keel of ice ridge, <*ca.* 25 m	Ice-strengthened steel jackets	Fixed production units or use of subsea wells if these cannot be damaged by drifting ice	Fixed production units. Any subsea wells must be trenched to avoid damage from drifting ice ridges	Caisson type of ice-resistant structures
Intermediate, <*ca.* 50 m	Ice-strengthened steel jackets. Beware of dynamic effects caused by breaking ice	Fixed production units with additional subsea wells in "excavated holes" tied back to fixed platforms	Fixed production units with possible tie-back of subsea wells. Full wellstream if feasible, possibly with a surface support vessel	Fixed production units with possible tie-back of subsea wells. Full wellstream if feasible, possibly with a surface support vessel
Deeper Arctic waters	Traditional deepwater concepts capable of managing the impacts from ice loads	Disconnectable vessels in combination with subsea wells. Ice feature management required. Full wellstream if feasible, possibly with a surface support vessel	Disconnectable vessels in combination with subsea wells.Ice feature management required. Full wellstream if feasible, possibly in connection with a surface support vessel	Full wellstream to shore. Note that the drilling season may be very short

After disconnecting production facilities due to ice restrictions, it may take a long time to reconnect.

Deepwater Arctic gas/condensate fields tend to be several hundred kilometers from shore, in 300–350 m water depth, and exposed to occasional ice and drifting icebergs. Due to the distance from shore, such a development may be unsuited to a full wellstream transfer to shore. It may be advantageous to use support vessel(s) at the site for control, MEG injection, and possibly separation purposes and compression.

Table 7.1 summarizes some suggested applications of subsea concepts for different water depths and ice conditions in the Arctic.

7.5 Deepwater technology

Exploration and production activities have increasingly turned to deepwater prospects, for example, in more than 500 m of water depth. These hydrocarbon accumulations have previously been out of reach for conventional platform developments, but the use of floating production facilities and subsea technology has made exploitation in such areas possible. Such water depth presents major challenges to production. It dominates all process, design, and economic considerations (Amin et al., 2005). Oil production takes place at water depths of 3000 m offshore Brazil. The major deepwater areas in the world are offshore Brazil, West Africa, and in the Gulf of Mexico.

In several deepwater areas currently under consideration for developments, there are mud diapers on the sea bottom and a potential for instability of the sea bottom. The sea bottom outside the large river deltas is likely to consist of very soft materials. The constant transport of mud with large rivers may cause mudflows or seafloor slides. These conditions will influence the possible location of wells, templates, and pipeline routing.

For deepwater developments, a number of technological challenges remain to be solved or improved upon. Production technology used for water depths up to 500 m can generally be extended to 1500 m and beyond. The technology for laying and repair of large submarine pipelines has been studied for an Oman to India gas pipeline system, and a trans-Black Sea gas pipeline is in operation (2000 m water depth).

7.6 Summary

Onshore developments feature considerable civil construction works of large, heavy, space-consuming structures, whereas development of offshore fields requires different development strategies. The costs of platform support structures increase considerably with increasing topside weight, and it is therefore important to reduce the amount of processing equipment offshore where possible. Offshore developments usually have fewer wells than corresponding onshore reservoirs, and require careful optimization of the well pattern.

Offshore development concepts include fixed platforms, floating platforms, subsea wells, and ship-shaped units such as FPSOs. Complete subsea development with tie-back to onshore facilities is possible and relevant in some cases. The buoyancy of any body submerged in water and the stability requirement to vessels and the effects of waves and currents must be taken into account.

As offshore oil and gas regions mature, the focus of exploration and development shifts toward smaller hydrocarbon accumulations, which are in many cases better suited to subsea development solutions than to the more costly stand-alone platform developments.

Subsea systems comprise of the well system (includes the downhole completion system and the subsea tree), the production system (includes protective structures, manifolds, templates, intervention systems, and subsea processing systems), and the pipeline system (includes tie-ins, umbilicals, risers, injection pipelines, and production pipelines).

There has been an increasing focus on deepwater prospects that have previously been out of reach. The use of floating production facilities and subsea technology has made exploitation of such fields possible, but a number of technical challenges remain.

References

Amin, A., Riding, M., Shepler, R., Smedstad, E. & Ratulowski, J. 2005. Sub sea development from pore to process. *Schlumberger Oilfield Review*, Spring 2005, 17(1).

Christie, A., Kishino, A., Cromb, J., Hensley, R., Kent, E., McBeath, B., Stewart, H., Vidal, A. & Koot, L. 1999. Sub sea solutions. *Schlumberger Oilfield Review*, 11(4) pp. 4–17.

Jahn, F., Cook, M. & Graham, M. 1998. *Hydrocarbon Exploration and Production*. Elsevier B.V., Amsterdam.

The project development process

Planning, deciding, and implementing a large oil and gas development is a major managerial challenge. Petroleum projects are usually large in terms of their complexity, the sums of money involved, and the potential for negative consequences of failure.

Various companies in the industry have developed their own models, principles, and procedures for carrying out large projects. The review in this chapter is based mainly on a methodology developed by Statoil, on the basis of experience from many large projects primarily in the Norwegian offshore. This experience includes large successes as well as problems of delays and cost overruns, while often moving at the very forefront of global technology development for challenging offshore production environments as well as safety and environmental standards.

8.1 Project development overview: from discovery to production

Exploration and Production (E&P) ventures comprise stages of activities, starting with screening of potential exploration areas and ending with field abandonment. Along the development there are a number of decisive events, from license award to end of production, which define the stages. Figure 8.1 illustrates the entire life cycle of an E&P venture in stages.

This section will be concerned with two of those stages: appraisal and development planning, and project execution. They can jointly be referred to as the project development. These are the stages in which the oil firm makes large commitments of resources toward

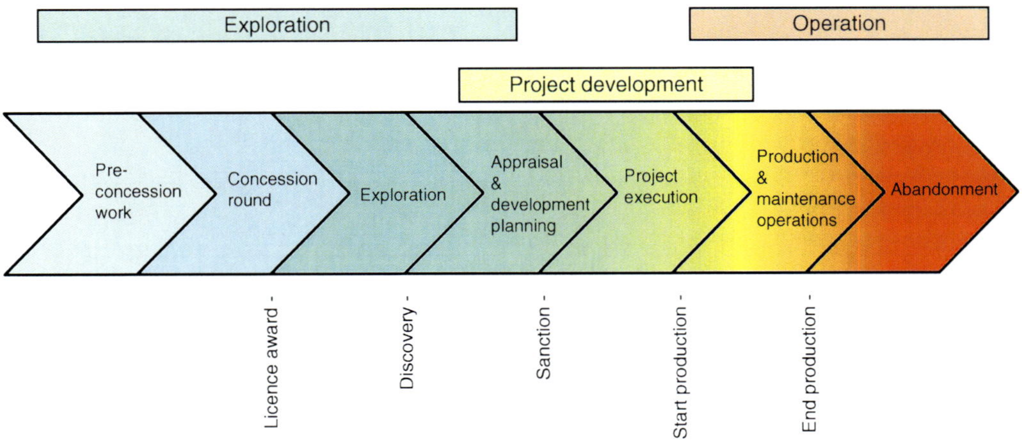

Figure 8.1: Stages of an Exploration and Production venture.

a petroleum development. These stages start with petroleum reserves having been identified, and end with the completion of installations ready to produce (or end with a decision not to develop). Clarifying the basis for making such commitments, selecting the best available alternative, and aligning contributors toward an overall goal are main challenges of this process.

8.1.1 Project development phases and decision gates

Before project development is initiated, the potential project is essentially a business idea, for instance, to exploit hydrocarbon resources that have been identified by exploration. Project development comprises two major stages: project planning and project execution (project planning corresponds to appraisal and development planning in Figure 8.1).

The project planning stage has the purpose of developing the business idea to a point where a thoroughly prepared decision can be made to develop the project (project sanction), or alternatively to reject the project as not viable at this time. The project execution stage has the purpose of making all practical and formal preparations for start of production.

The two major stages of project development are divided further into *phases*, each of which has a defined purpose and defined products. *Decision gates* (DG) are events which mark important decisions as well as the transfer from one phase to the next.

The decision-making process applied to a project must be consistent with the regulatory requirements applicable in the jurisdiction (country) where the project is located. This may require certain adaptations to the process relative to what is presented here.

The concept of DG is used as a tool for maintaining control of the project and to ensure a structured decision process. They constitute "milestones" where the status of the project is reviewed in order to conclude whether the project will be continued, whether major changes need to be made, or whether the project should be terminated.

Following is a brief overview of the main purposes of the phases, DG, and approval points of the project development process as illustrated in Figure 8.2.

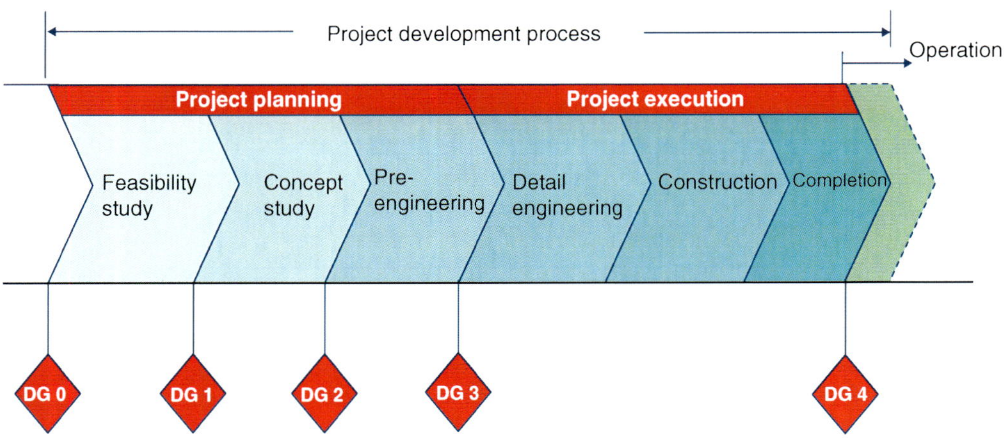

Figure 8.2: The project development model for investment projects.

8.1.1.1 Phases of the project planning stage

Feasibility: Establish whether hydrocarbon resources (or other business opportunity) are technically and economically feasible. Identify one viable development concept, though not necessarily the best development concept.

Concept: Select one development concept after identifying and evaluating alternatives. Confirm feasibility and profitability of the selected concept. In some cases it may be necessary to still pursue more than one alternative concept beyond the concept phase.

Pre-engineering: Develop and document the business opportunity to such a level that a final project sanction can be made, application to authorities can be sent, and contracts can be entered into. This phase is also referred to as front end engineering and design (FEED).

8.1.1.2 Phases of the project execution stage

Detail engineering: Prepare detailed technical plans and drawings of the installations as a basis for construction. Enter into main contracts for procurement and construction.

Construction: Construct and install the installations. Drill the wells that will be used from start-up.

Completion: Commission the installations to be ready for production. Complete preparations for production operations and confirm their readiness for safe operations.

8.1.1.3 Decision gates

DG imply an internal commitment of resources necessary to proceed to the next DG. At DG 0, 1, and 2, the resources committed are those required for planning and appraisal activities, whereas DG 3 commits the major resources required to fully develop the project.

DG 0: Decision to initiate the feasibility phase, on the basis of a business idea and identified petroleum resources.

DG 1: Decision to enter the concept phase, on the basis of identified feasibility of a development.

DG 2: Decision to undertake pre-engineering, on the basis of one (or more) selected concepts for the development.

DG 3: Project sanction, that is, decision to carry out the development based on a development plan and budget.

DG 4: Decision to start operating the installations, which have been confirmed as ready for safe operations.

Although the process is divided into phases, the project is managed as an integrated process from the feasibility phase to start-up of regular operation. The organization and control system will be developed and personnel may change during the course of the project, but a high degree of continuity must be aimed for.

8.2 The planning stage (DG 0 $\rightarrow$ DG 3)

The planning period covers the feasibility, concept, and pre-engineering phases. It will result in either the project being ready for sanction (approval of the development) or the project being rejected as not viable. The planning period requires a systematic screening of a range of concepts, and definition of a selected concept, to clarify whether a business opportunity can be developed to satisfy the company's requirements for profitability, HSE ((protection of) health, safety, and environment), and technical feasibility within defined limits of uncertainty. To reach project sanction, the planning stage must also yield a development plan and a budget as basis for project execution which follows.

8.2.1 Decision gate 0: decision to initiate the feasibility phase

A decision to initiate a feasibility study for a potential development can be made on the basis of a promising indication that an attractive project may be developed, with a need for further investigation before an investment decision can be made. One or more exploration wells that have encountered and tested hydrocarbons, in an area where the firm can have the right to exploit hydrocarbons, would be such an indication. In other cases, the decision to initiate a feasibility study may come from a business idea not linked to a new discovery, for instance, concerning modifications or redevelopment which could significantly improve the economics of a previously developed field.

At DG 0, a project agreement should be established. It is not necessarily an agreement in the legal sense, but an outline of further work which is agreed to by those in the organization who will be responsible for carrying out the work, and by those who have the authority to commit resources to it. The latter may include other oil firms which are partners in the license area where the potential development is located.

A project agreement should outline the following issues, and should be updated as necessary at later times in the project:

- definition of project goal (what to be accomplished by the feasibility study);
- the tasks to be carried out;
- budget and manpower;
- other conditions and assumptions.

8.2.2 Requirements for the feasibility phase (DG 0 $\longrightarrow$ DG 1)

The main purpose of the feasibility phase is to determine and document whether a business opportunity is technically feasible and has an economic potential to justify a development. The feasibility phase is initiated at DG 0, and leads to DG 1, which is a decision to start concept development.

A key component of a feasibility study is to describe one possible concept for developing a project (i.e., one possible development solution), and to demonstrate that this would be a viable project in terms of technology, HSE requirements, and economics based on reasonable assumptions that can be made. The concept thus described would not necessarily be the best possible concept (as this would be a matter for the next phase to determine), but may serve as a reference against which other possible concepts are compared in the next phase.

The following activities are required during the feasibility phase:

1. Project management:

- Appoint a person responsible for executing the project (project manager), as well as other members of the project organization with specified responsibilities. The project organization can be a dedicated organization or a matrix organization. In the latter case, the project organization consists of members who at the same time continue to have their regular responsibilities reporting to managers outside the project.

- Update the project agreement as may be required.
- Establish agreements with partners required for the feasibility phase.
- Establish a list of agreements required for further development of the project.
- Establish objectives for the concept phase.
- Benchmark key project parameters against comparable projects.
- Assess project status against DG 1 requirements.

2. *Project concept*:

- Review, evaluate, and document the basis for the business opportunity.
- Describe the valuable product or other output that will generate income for the project, including an evaluation of potential markets.

3. *Technical development*:

- Prepare a preliminary design basis document.
- Describe one facility concept as a reference case, and document its technical feasibility.
- Identify other concepts which can be possible, including any concepts on the basis of new technology which have a potential for improving the project.
- If deemed to be critical, carry out a screening of possible technologies and potential supplier(s).
- Tentatively assess local technical and operational requirements.
- Assess the suitability of available technology for the requirements of the reference case.
- Develop a technology qualification program (a plan for verifying available technology for the requirements of the project).
- Assess the requirements for drilling activities (if applicable).
- Establish a regularity management program.

4. *HSE*:

- Identify HSE challenges, hazards, and relevant authorities' requirements. Evaluate associated risks and identify risk-reducing measures.

5. *Commercial issues and project economy*:

- Tentatively analyze the profitability of the business opportunity in accordance with corporate guidelines.
- Identify the potential impacts of contingent (uncertain) outcomes.
- Evaluate the need for long-term sales and/or transport agreements to be established prior to project commitment (particular relevance for gas sales, oil sales by pipeline, and use of infrastructure owned by other parties).
- Identify any potential discrepancy of commercial interest between involved parties, that is, areas in which affected parties will be likely to prefer different outcomes.
- Evaluate requirements for commercial agreements or arrangements in relation to each of the identified technical solutions.

6. *Project control*:

- Establish a cost estimate with a relatively wide range of uncertainty, for example, $\pm 40\%$.
- Review the sources of uncertainty for the project, including the resource basis, market, technical solutions, HSE requirements, project execution, vendor market, cost estimate, and profitability (as relevant).

- Establish a mitigation plan to reduce uncertainties (outlining risks and opportunities).
- Establish a work program, plan, budget, organization, and reporting system for the concept phase activities, which will follow DG 1.

The outputs from the feasibility phase constitute the documented decision basis for passing DG 1, and form the basis for the concept phase.

8.2.3 Decision gate 1: decision to start concept development

The DG 1 approval is an authorization by the company and the partners to continue developing the project through the concept phase toward DG 2 in accordance with the approved project plans and budgets.

The basis for such authorization must be that the company management and partners are satisfied on the basis of the feasibility work that opportunities for a viable and attractive project exist. In some cases there is not sufficient basis to pass DG 1, in which case the project manager reports and recommends the project to be discontinued (or the project manager's recommendation to proceed to concept phase is turned down).

8.2.3.1 Timing

DG 1 may be passed when the business concept has been developed to a level where it is demonstrated as likely to be technically feasible, profitable, and consistent with corporate business plans.

8.2.3.2 Documentation

The DG 1 decision requires a memorandum and supporting document that reviews the business opportunity and the development of the project up to DG 1, and concludes that the DG 1 requirements are met. Any deviations from governing documents must be identified. For modification projects, a recommendation on whether to use this procedure as the basis for further development of the project shall be included.

8.2.3.3 Quality control

Documented results of the quality control process are required at DG 1 in accordance with company procedures.

8.2.3.4 Recommendation and approval

The DG 1 proposal should be reviewed by the functional management for exploration and for project development, and their recommendations included in the documentation presented for DG 1 approval.

The company system for delegation of authority determines how DG 1 can be approved within the company. Approval by project partners (co-owners) is also required in accordance with applicable agreements for the partnership.

8.2.4 Requirements for the concept phase (DG 1 $\rightarrow$ DG 2)

The concept phase should normally result in one concept (a development solution) being recommended and documented as feasible, after identifying and assessing all other relevant alternatives. The design basis is defined in terms of resources and product. The profitability and feasibility of the chosen concept should be documented in accordance with company guidelines and business plans. The selected concept is recommended for DG 2, "provisional project sanction."

The following activities are required during the concept phase:

1. Project management:

- Establish a documented management system suited to the project scope and size.
- Make any required adjustments to the project organization.
- Establish goals and targets for the project, covering profitability, regularity, project execution, HSE, and quality assurance.
- Update the project agreement as required.
- Carry out a stakeholder analysis appropriate to the project's scope, complexity, and other requirements.
- Develop and document a project execution strategy.
- Based on the project execution strategy, develop strategies for:
  - commercial agreements;
  - information technology;
  - operation and maintenance.
- Develop a project execution plan (PEP) describing the project and the management system. It will be based on the project agreement, the project execution strategy, and the overall procurement strategy, and should be developed and updated regularly during the project planning period.
- Establish a change control system, referring to an established basis.
- Establish goals for the pre-engineering phase.
- Benchmark the project against comparable projects.
- Measure project status against DG 2 requirements.

2. Project concept:

- Review, evaluate, and document the basis for the business opportunity.
- Update the product description and market analysis as required.

3. Technical development:

- Review and update the design basis document, including infrastructure evaluations where relevant.
- Assess reports from reviews and verifications.
- Assess the requirements for drilling activities and equipment (if applicable).
- Develop a production strategy for the field (upstream).
- Update the regularity management program and carry out the activities required by it.
- Carry out a regularity analysis for the total production system.
- For concepts that require ship transportation, carry out a shipping simulation study.
- Carry out an operation verification of the design.
- Identify and evaluate all relevant concept alternatives and concept variants.
- Propose and select the best concept solution(s).
- Identify and pursue opportunities for improving the project value.
- Define the selected concept(s) in accordance with the requirements for the cost estimate and its range of uncertainty.
- Establish a technology qualification program if relevant.
- Establish preliminary project technical and operational requirements and guidelines.
- Implement a documentation system for handling technical information.

4. HSE:

- Identify challenges and hazards with regard to health, working environment, safety, security, and environment. Assess the risks, and identify risk-reducing measures.
- Identify relevant regulatory HSE requirements and plan for compliance.
- Establish a plan for preparation of the environmental impact assessment (EIA), to ensure that the EIA process can be completed within the framework of the project main schedule. Depending on national legislation, the plan for preparation of the EIA may need to be submitted to a public review before the EIA itself is prepared.
- Perform a total risk analysis.
- Establish a HSE program and plan.

5. Procurement:

- Establish a procurement strategy and award contracts covering procurement needs in the concept phase.
- Develop an overall procurement strategy consistent with the project execution strategy. The procurement strategy shall include descriptions of supplier market, contract packages, purchasing strategy, identification of long lead items, and use of frame agreements.
- Prepare a specific strategy and tender invitations for pre-engineering contracts.

6. Commercial issues and project economy:

- Carry out profitability analyses to assess whether the business opportunity meets corporate requirements for profitability. The analyses shall include value chain analyses through to the end customer.
- Identify and describe all financial and commercial agreements and arrangements that are relevant to the project development process. Establish a strategy and a plan for entering into agreements.

7. Project control:

- Establish a planning system with a main plan detailing main project activities, milestones, involvement with authorities and partners, and main supervision activities.
- Develop cost estimates to a narrower range of uncertainty than in the feasibility phase, for example, $\pm 30\%$.
- Carry out a comprehensive uncertainty analysis covering all relevant technical and commercial aspects, that is, resource basis, market, technical solutions, HSE, project execution, supplier market, cost estimates, and resulting profitability.
- Update mitigation plan to reduce uncertainties (outlining risks and opportunities).
- Propose a plan, budget, and organization for the pre-engineering phase after DG 2.

The outputs from the concept phase constitute the documented decision basis for passing DG 2, and form the basis for the pre-engineering phase.

During the concept phase, one concept or, where necessary, a limited number of concepts are chosen for further detailing toward DG 2. This requires a screening process including all relevant and feasible alternative concepts identified for a further development of the business opportunity. Circumstances may dictate that more than one concept is selected for further development, or that a concept is selected at another point in time in the planning period. The selection of concept(s) should be recommended by managers of affected departments and functions.

The selection of the base case concept(s) requires supporting documentation describing the concept screening process, focusing on:

- the design basis;
- concept alternatives and variants that have been assessed;
- screening criteria and weighting;
- description of and justification for both the selected concept(s) and the rejected options;
- a technology qualification program to verify the suitability of the recommended technology, if required.

8.2.5 Decision gate 2: provisional project sanction

The DG 2 approval is an authorization by the company and the partners to continue developing the project through the pre-engineering phase toward DG 3. The approval should include project plans and budgets for the pre-engineering phase.

DG 2 approval is also a decision to develop and submit the necessary planning applications to the authorities, in accordance with applicable regulations. For petroleum field developments in Norway, the planning application is referred to as a Plan for Development and Operations (PDO), to be supported by an EIA. Other jurisdictions may have similar requirements.

8.2.5.1 Timing

DG 2 may be passed when the business concept has been developed to a level where it has been documented as feasible, profitable, and consistent with corporate business plans.

8.2.5.2 Documentation

The DG 2 decision requires a memorandum and supporting document that reviews the business opportunity and the development of the project up to DG 2, and concludes that the DG 2 requirements are met. Any deviations from governing documents must be identified.

The DG 2 documentation should include an evaluation of the availability of qualified personnel resources and of the capacity in the relevant supplier industry.

8.2.5.3 Quality control

Documented results of the quality control process are required at DG 2 in accordance with company procedures.

8.2.5.4 Recommendation and approval

The DG 2 proposal should be reviewed by the functional management for exploration and for project development, and their recommendations included in the documentation presented for DG 2 approval.

The company system for delegation of authority determines how DG 2 can be approved within the company. Approval by project partners (co-owners) is also required in accordance with applicable agreements for the partnership.

8.2.6 Requirements for the pre-engineering phase (DG 2 $\rightarrow$ DG 3)

The purpose of the pre-engineering phase is to further develop and document the business opportunity based on the selected concept(s) to such a level that a final project sanction can be made, application to authorities can be sent, and contracts can be entered into. The pre-engineering phase leads to DG 3 "project sanction."

The following activities are required during the pre-engineering phase, many of which are continuations of activities described above for the concept phase:

1. Project management:

- Make any required adjustments to the project organization and the documented management system.
- Update all project plans and strategies as may be required. This includes the project goals and targets, the stakeholder analysis, the project agreement, execution strategy, strategies for commercial agreements and information technology, and the PEP.
- Develop a commissioning strategy as a basis for contracts, engineering, and construction planning.
- Implement the change control system for the execution period.
- Prepare the required applications to the authorities.
- Benchmark the project against comparable projects.
- Measure project status against DG 3 requirements.

2. Project concept:

- Review, evaluate, and document the basis for the business opportunity.
- Update the product description and market analysis as required.

3. Technical development:

- Review, update, and finalize the design basis document, so as to require no further changes during the project.
- Assess reports from reviews and verifications.
- Optimize the selected concept option(s).
- Define the selected concept in accordance with the requirements for the cost estimate and its range of uncertainty.
- Update the technology qualification program if required.
- Complete and approve the project's technical and operational requirements.
- Enter into license agreements for the selected technology and processes as required for licensed technologies.
- Complete drilling plans and drilling equipment requirements (if applicable).
- Update the production strategy for the field (upstream).
- Update the regularity management program and carry out the activities required by it.
- Update the regularity study for the total production system.
- Update the shipping simulation study if applicable.
- Update the operational verification of the design.
- Develop the operation and maintenance strategy for the facility.
- Involve selected suppliers in the above processes as required in accordance with agreements and the overall procurement strategy.

4. HSE:

- The EIA program shall be established and approved, and the necessary study work carried out.
- The total risk analysis shall be updated as necessary.
- The HSE program for the execution period shall be completed.

5. Procurement:

- Award pre-engineering contract(s).
- Update the overall procurement strategy as necessary, contingent on DG3 approval.
- Place purchase orders for long lead items as required by the contract plan.
- Develop specific procurement strategies for the execution phase.
- Prepare tender invitation documents as required.
- Develop a contract plan and basis for entering into contracts and purchase orders.
- Place contracts and purchase orders in accordance with the approved contract plan.

6. Commercial/economy:

- Update and confirm economic analyses and profitability calculations. For the project to pass DG 3, the economic analyses must demonstrate that it meets the company's requirements for profitability.
- Update the strategy and plan for entering into financial and commercial agreements and arrangements. All critical agreements shall be established and approved before DG 3, but shall contain clauses making the agreements conditional on DG3 approval.

7. Project control:

- Develop the planning system into a main plan that shows the project's main activities, main milestones, important products, activities toward the authorities and partners, and main inspection activities.
- Establish a resource and manpower plan for the execution period.
- Establish a supervision plan.
- Establish cost estimates at a narrow range of uncertainty, for example, ±20%, and corresponding budget proposals.
- Establish a project control basis.
- Update the uncertainty analysis and mitigation plans covering the resource basis, product markets, technical solution, HSE, project execution, supplier market, cost estimate, and profitability.
- Evaluate the availability of qualified personnel resources in the company and of the capacity in the relevant supplier industry, for executing the project.

The products from the pre-engineering phase constitute the documented decision basis for passing DG 3 and form the basis for the project execution period.

8.2.7 Decision gate 3, "project sanction"

The DG 3 approval is the commitment by the company and the partners to developing the project. DG 3 includes the approval of plans and budget on which the development will be based.

DG 3 is also a decision to submit application to the authorities for the project as required by national legislation.

For projects within the jurisdiction of the Norwegian Petroleum Act, a PDO or a "Plan for Installation and Operation" (PIO) is required, covering field installations and pipeline systems, respectively. The Norwegian Petroleum Directorate has issued "guidelines for PDO and PIO" specifying the requirements for the plan. The PDO or PIO must be approved by the responsible business unit, corporate management, the board, and the partners, before it is submitted. When the partnership submits a PDO or PIO to the authorities, this represents a commitment by the partnership to carry out the project development. For projects in this category, completion of the PDO (or PIO) and DG 3 should therefore occur at the same time.

8.2.7.1 Timing

DG 3 may be passed when the business concept has been developed to a level where it has been documented that it meets the established requirements with regard to profitability, HSE, technical definition, cost estimate accuracy, and project execution uncertainty.

8.2.7.2 Documentation

The DG 3 decision basis is a memorandum and supporting document that reviews the business opportunity and the development of the project up to DG 3, and concludes that the DG 3 requirements are met. Any deviations from governing documents must be identified.

8.2.7.3 Quality control

Documented results of the quality control process are required at DG 3 in accordance with company procedures.

8.2.7.4 Recommendation and approval

The DG 3 proposal should be reviewed by the functional management for project development, and its recommendation included in the documentation presented for DG 3 approval.

The company system for delegation of authority determines how DG 3 can be approved within the company. Approval by project partners (co-owners) is also required in accordance with applicable agreements for the partnership.

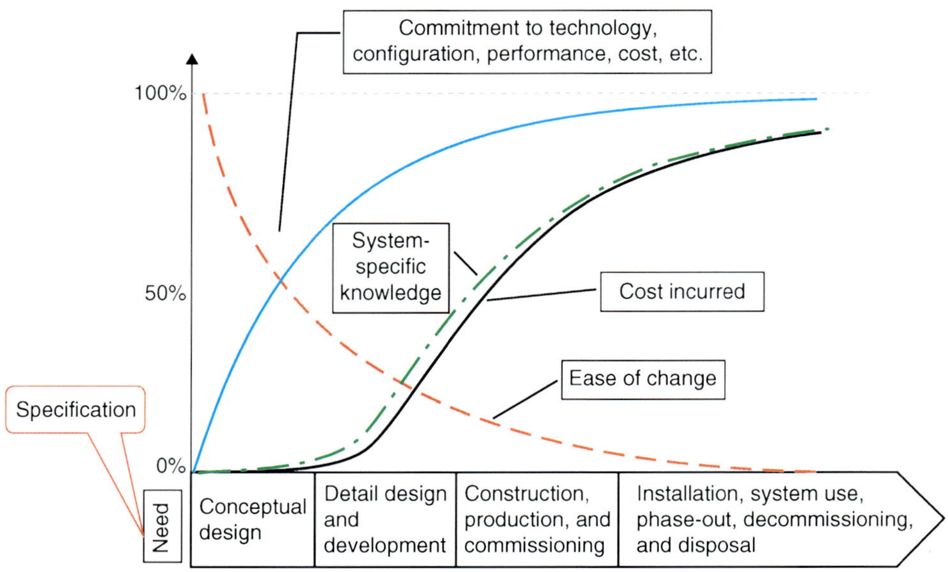

Figure 8.3: Project phases and commitment to cost and technical issues. (Conceptual design = project planning stage. The other phases are phases of the project execution stage, see Figure 8.2).

8.2.7.5 Experience transfer

After DG 3, a process of experience transfer should be provided for, which may be done as an experience transfer workshop, to the extent that the project execution will be managed by a different team than the previous phases.

8.2.8 Summary regarding project phases

Figure 8.3 summarizes the project phases and the commitment to cost and technical issues for a project that is being executed in accordance with the model presented in this chapter.

In the early phases, the cost incurred is low while the commitment to technology, configuration, performance, cost, etc., grows fast.

Decommissioning

Eventually the facilities installed for oil and gas production will reach the end of their economic lifetime. In a well-managed field development, this point will be reached only after having produced as much oil and gas as is technically and economically feasible on the basis of the best available techniques. Good oil field practices require, with very few exceptions, that the environment shall be restored to approximately its pre-exploration condition once oil and gas production has ceased. In some onshore environments, the oil firms and relevant authorities may agree that the production site be prepared for another use, such as for an industrial development.

Decommissioning, or abandonment, is the process by which the operator of an oil or gas installation will plan, gain approval for, and implement the removal, disposal, or re-use of an installation when it is no longer needed for its current purpose (Jahn et al., 1998). It applies not only to entire field developments at the end of their economic life, but also to individual installations such as wells that will no longer be used while other parts of the field are still producing.

9.1 Decommissioning: general considerations

Decommissioning is often a complex and costly operation which entails certain risks, particularly for offshore installations. In addition to the regulatory requirements, which will be reviewed in a later section, five key considerations influence the concept and method for decommissioning field installations (Jahn et al., 1998):

- the potential impact on the environment;
- the potential impact on human health and safety;
- technical feasibility;
- costs;
- public acceptability.

More than 1000 offshore installations have been decommissioned and removed to date, most of which were steel structures in the Gulf of Mexico. Some 7000 remaining offshore installations are located on the continental shelves of more than 50 seaside states. They include approximately 450 large structures, having a jacket weight of more than 4000 tons or located in water depth greater than 75 m. Removal of these large structures will be particularly difficult, costly, and potentially hazardous. In some cases, removal may not be possible with current technology. A rough estimate has indicated a total cost of US$ 40 billion for removing all offshore structures worldwide.

Increasing public awareness and regulatory requirements for environmental protection has increased the requirements for proper decommissioning and restoration of production sites. The practice of dumping disused equipment on the ocean floor has been common in the past, but is now widely considered as unacceptable or even illegal. In addition to the general environmental concern, such dumped items can destroy fishing equipment.

One remarkable event in that regard occurred in 1995, when the oil company Shell reversed its abandonment plan for the disused "SPAR" installation from the Brent field in the United Kingdom. The plan had been to float the installation to very deep Atlantic Ocean waters and dump it to the ocean floor. Permission to do so had been obtained from the UK government, but triggered strong protests from environmental interest groups with apparent support from large parts of the general public. Shell backed down from its original plan and instead sent the installation for dismantling and recycling in Norway, at a much higher cost to Shell but causing less displeasure with the company's many environmentally concerned customers and other stakeholders. Parts of the large concrete installation have now become foundations for a new jetty.

9.2 Decommissioning methods

9.2.1 Well abandonment

Abandoning a well, whether onshore or offshore, requires the following (Jahn et al., 1998):

- isolation of all hydrocarbon-bearing intervals;
- containment of all over-pressurized zones;
- protection of overlying aquifers;
- removal of wellhead equipment.

An abandonment process begins with "killing" the well, a process in which the produced fluids are circulated out of the well, pushed back into the formation, and replaced by fluids that are sufficiently heavy to contain any open formation pressure. Once the well is killed, the tree is removed and replaced by a blow-out preventer. The production tubing may be removed through the blow-out preventer (Jahn et al., 1998).

The perforations are then cemented to seal off all production zones. It is normal to set a series of cement plugs in both the production liner and the production casing. The production casing is cut and removed above the top cement, and another cement plug over the casing stub will isolate the annulus and any formation that may still be open below (Jahn et al., 1998).

9.2.2 Decommissioning offshore fixed platforms

Fixed offshore platforms can be removed completely or incompletely.

9.2.2.1 Complete removal

Platforms that are removed may be recycled or disposed of at shore, disposed of at deep water, or used as artificial reefs. In the Gulf of Mexico, where the largest number of platforms has been decommissioned to date, a majority of platforms were removed in the same manner in which they were installed as steel jackets, only in reverse. Most decommissioned platforms have been taken to shore. A particular problem arises where oil-based drilling mud has been dumped on the seafloor, whereby the disturbance caused by the removal process may lead to pollution release.

Concrete platforms will be more difficult to remove. There are relatively few (but some very large) concrete platforms worldwide. They were originally floated out to location using tanks in the substructure for buoyancy, but tend to settle into the seafloor on location, making them difficult to refloat. The disused concrete platforms at the Norwegian Frigg field have been left on location indefinitely, after decommissioning the wells and removing certain items.

9.2.2.2 Incomplete removal

Incomplete removal of a platform can involve toppling, partial removal, leaving in place, or re-use. These options may be limited by applicable legislation and environmental concerns.

Toppling an installation involves cutting off the legs a little above the mud line. Then it tips over and remains on the sea bottom. Some steel structures in the US Gulf of Mexico have been toppled so as to create artificial reefs, which may enhance marine life.

With *partial removal*, all or parts of structures are left on the seabed at or near the original location of the structure. The entire deck and a portion of the supporting structure are removed down to a depth that allows for free navigation.

Leaving an installation in place can be a default option particularly if no viable method for removal is found, which may be the case for large concrete support platforms. It would still require proper abandonment of wells, cleaning, and removal of items on the deck that can be moved.

Re-using platforms for a different purpose after refurbishment has taken place in several countries – most common in the US Gulf of Mexico. It has also been done in the North Sea and Southeast Asia. Re-using an existing asset defers its final disposal.

9.2.3 Decommissioning floating, subsea, and land installations

Floating platforms such as semi-submersibles, tension leg platforms (TLPs), and ship-shaped units can easily be released from their mooring and be towed away for service elsewhere. This is a cheap and attractive option. Ultimately they are broken up at ship-breaking facilities and recycled as scrap metal.

Figure 9.1: Lifting of the Tommeliten subsea template. Photo: Stein Are Osnes/Statoil.

Subsea facilities can usually be removed with relative ease as they are small, modularized, and easy to lift. However, manifolds and templates may weigh in excess of 1000 tons and will thus require heavy crane barges for removal. Figure 9.1 shows the removal of a subsea template at the abandoned Tommeliten field in the Norwegian offshore.

Pipelines will be circulated clean, and those that are buried or left on the seabed will be filled with water or cement. Flexible subsea pipelines may be reeled in and disposed of onshore. Surface piping will be removed. A steel pipeline can have a long useful lifetime, usually more than 50 years if constructed and maintained well, and it can be relevant to leave it in place for possible future use.

Land facilities such as processing facilities and terminals will be cleaned and scrapped. The land under the facilities may have to be reconditioned if pollutants have escaped during operations. The return of industrial sites to green field conditions have proved to be very expensive for companies in the USA, with several law suits outstanding brought on by local groups and environmental groups.

In 2000, Shell closed down an oil refinery outside Stavanger, Norway. The company subsequently spent close to $80 million to carefully restore the ground, including the use of steam to clean the soil of pollution that had seeped from the refinery. The area has since been redeveloped as a harbor and industrial site. This operation by Shell contrasts sharply with many oil sites worldwide which have been abandoned with high and enduring contamination of the soil.

9.3 Legislation

The coastal state has rights as well as obligations under international law. One such obligation is to secure that disused or abandoned equipment is removed from the seabed.

9.3.1 United Nations conventions and rules

In order to protect other users of the sea, and to protect living species and their harvesting, the coastal state is obliged to clean the seabed from harmful or useless remains of the petroleum activity. A general provision to this effect is found in the UN Convention on the Law of the Sea, Article 60, No. 3.

In addition, international guidelines for the protection of navigation apply to the abandonment of installations, with their origin in Article 60 of the Convention of the Sea. The competent international organization is the marine arm of the UN: the International Maritime Organization (IMO). The IMO rules and guidelines on removal of installations were adopted on October 19, 1989. Those guidelines permit partial removal of offshore installations, but only when particular circumstances are present. The IMO rules do not relate to pipelines, and they do not apply in the territorial waters of any state.

The main rule in the IMO document is that full removal is required if the installation is erected in waters shallower than 75 m, and the weight of the same installation is less than 4000 tons. Of the world total of more than 7000 platforms, the IMO rules will lead to complete removal of more than 95% of them.

9.3.2 The London Convention

The Convention on the Prevention of Marine Pollution by Dumping of Wastes and Other Matter, commonly known as the London Convention of 1972, has been signed by a large number of coastal states. It concerns the protection of the seas from pollution. The Convention prohibits dumping of platforms without prior permission by the relevant coastal state. Harmful substances must under any circumstance be removed from the structure before dumping. The leaving of a platform in place with no other purpose than disposal through breakdown is defined as dumping in the terms of the Convention.

The London Convention has been supplemented by a London Protocol, which entered into force in 2006, with tighter requirements in some regards.

9.3.3 The OSPAR Convention

The Oslo–Paris (OSPAR) Convention concerns the protection of the marine environment in the northeast Atlantic Ocean, and was signed in Paris in 1992. The Annex on Offshore Activities deals with disposal of disused installations. Dumping or leaving disused installations in the ocean is basically prohibited, but can be permitted in accordance with certain procedures if there are strong reasons for preferring this alternative. The prohibition does not apply to pipelines and certain other equipments on or beneath the seafloor which do not obstruct fishing.

Following a Ministerial Declaration, governments are encouraged to ensure that offshore installations deployed after January 1, 1998 are designed in a way that does not exclude any disposal alternative, that is, it must be feasible to remove such installations.

9.3.4 National legislation

Such international conventions as mentioned above generally constitute obligations by the signatory nations to implement corresponding rules in their national legislations.

In Norway's case, the international obligations are followed up in the Petroleum Act, Chapter 4. It obliges the licensee to submit a decommissioning plan before expiry of a license or the permanent termination of use of a facility. The state may then elect to take over the title to all installations free of charge. In that case, the obligation to remove is transferred to the state. If the state does not exercise its right to take over the installations, the obligation to remove remains with the owner. In that case, the ministry may require the owners to remove the installation. There are certain provisions for sharing the costs of this between the state and company on the basis of taxes that the company has paid while producing from the field.

The right for the state to take over facilities does not include moveable units (whether floating or not). Neither does it include equipment owned by third parties.

9.4 The economics of late-life field management and decommissioning

The cost of removal can be very high. A medium-sized single steel jacket with its topside may cost about US$ 60 million to remove. Larger platforms may cost many times this amount, if they can possibly be removed at all.

Oil and gas fields tend to reach a plateau level of production, which can be maintained for a number of years, after which production gradually decreases. Declining oil and gas production, increased water production in many cases, and increasing maintenance costs of aging facilities erode the economics of production. Simply put, production from a field should be discontinued and the facilities decommissioned when production has declined to a point when its value no longer exceeds the expenses of continued operations, and all viable options for improving production have been exhausted.

The lifetime of a field may be extended by (from Jahn et al., 1998):

- reducing the operating costs;
- increasing the hydrocarbon production;
- increased price for hydrocarbons in the market.

Operating costs represent the major expenditures late in oil or gas field's life, and are determined by the number of staff required to run the facility, hardware and maintenance expenses, oil and gas processing expenses, needs for well interventions, etc.

Enhanced recovery processes such as improved oil recovery (IOR) should be considered as a means of deferring decommissioning by increasing the hydrocarbon production. Such methods are often very sensitive to the oil price, and while common for onshore developments, they are harder to justify offshore.

In some cases, companies may be motivated to produce beyond the point when production value no longer meets expenses. There are two reasons for this: one is the advantage of deferring decommissioning costs into the future, since the obligation to decommission usually arises when production stops permanently. Decommissioning will often be a large expense item, the economic burden of which is reduced by delaying it (the economic benefit of deferring expenditure is explained in Chapter 13). The other reason for continued production concerns option value: a company may hang on to production from a marginal field in the hope that things will improve, for instance, from increasing oil prices or a new discovery that can use the installation.

In many cases, oil fields change ownership in their late life. Certain organizations specialize in extracting value from small operations in mature fields, and may buy such fields from a previous owner whose organization is better suited to large-sized operations.

9.5 Funding the costs of decommissioning

The large costs of decommissioning occur at a point where the development is no longer generating any surplus funds. The decommissioning therefore needs to be funded by money previously generated from the field and set aside for decommissioning, or by other funds. The accounting rules applied by many oil companies internationally stipulate that the owner of a producing field must account for the estimated future decommissioning costs as a liability.

Raising funds for removal may sometimes be an economic as well as a legal challenge. When the petroleum deposit is depleted, the companies may find themselves with no cash flow to pay for removal expenses. From the government's viewpoint, it is necessary to obtain security that the companies will indeed fulfill their obligations for proper decommissioning, particularly with small companies which might be tempted to withdraw funds from their national affiliate so as to leave it financially insolvent. A similar concern would arise within a partnership of companies that share the obligation to finance the decommissioning work.

An important aspect of the removal issue is the tax treatment of removal costs. This challenge arises because the companies will incur decommissioning costs as a large expense item while not having any current income against which to match it for tax purposes. The solution in the United Kingdom is to allow full deduction of removal cost against other income the year it materializes, and a limited right to carry back expenses against the previous 3 years of income.

The Norwegian solution has been to let the state participate directly in the cost of removal through a grant. The share of removal cost carried by the state shall correspond to the average tax paid during operation on income from the installation in question. The intention is to secure cost- and tax-efficient removal and at the same time avoid the accumulation of untaxed funds in the companies' books. The uncertainties as to both when and how removal will take place have rendered it undesirable for the state to allow the operator to accumulate untaxed funds to meet this kind of future expenses.

9.6 Summary

Eventually every field development will reach the end of its economic lifetime. When the viable option for extending the operational life of the field has been exhausted, decommissioning will be required by law as well as by environmentally responsible oil field standards. Decommissioning is the process by which the operator of an oil or gas installation will plan, gain approval, and implement the removal, disposal, or re-use of an installation when it is no longer needed for its current purpose.

Decommissioning includes both the killing of the wells and abandonment of the facilities. Killing the wells involves sealing off the pathway from the formation to the surface. Abandonment of the facilities involves removal or partial removal of onshore structures, offshore platforms, subsea installation, pipelines, etc.

Among others, the UN Convention on the Law of the Sea, Article 60, No. 3, describes the coastal state's obligations to secure that disused or abandoned equipment is removed from the seabed, that users of the sea are protected, that living species and their harvesting are protected, and that the seabed is cleaned from harmful or useless remains of the petroleum activity.

Raising funds for decommissioning presents economic as well as legal challenges, to which various jurisdictions have selected different solutions.

Reference

Jahn, F., Cook, M. & Graham, M. 1998. *Hydrocarbon Exploration and Production.* Elsevier B.V., Amsterdam.

Safety management

Safety management is the application of managerial principles and techniques for the purpose of safeguarding people, the environment, and assets from accidents.

Besides the managerial aspects of safety, which is the main topic of this chapter, safe hydrocarbon operations also crucially depend on prudent technical control and safety systems, notably:

- process control facilities, whose main function is to assure that all production and utility systems operate within the prescribed design and alarm limits;
- process and emergency shutdown systems, designed to prevent hazardous situations from occurring (such as fluid levels, pressures, or temperatures out of control);
- fire and gas systems, for detecting fires and leakages of gas;
- condition monitoring systems, to monitor the integrity of structures, valves, rotating equipment, and other equipment.

Decision makers can have access to large amounts of relevant information in the area of risk and safety. This information must be made available and properly assessed. However, safety issues are often subject to uncertainties which must be taken into account.

Safety management also concerns attitudes, motivation, and ethical decision making. The attitudes of an organization's leaders to safety matters and their ability to motivate the entire company staff are crucial to successful safety management. Involving relevant personnel in risk-related decisions also improves the attitudes and ensures that the risks are known to those affected.

10.1 Responsibilities and principles for safety management

In a petroleum project or operation, responsibility for safety cannot be separated from the responsibility for overall results. Therefore, the responsibility for the total safety rests with the chief officer (top manager) for the project or operation. The chief officer cannot evade this responsibility by delegating it to other personnel. It is often useful to employ a safety engineer who can assist and carry out various tasks in safety management as directed by the chief officer. However, the chief officer must be directly involved in safety management and take responsibility for the final product. The responsibilities can then be defined as follows:

- Safety management is directed by the chief officer having overall responsibility for the project or operation.
- Each member of the project or operational team is responsible for safety within his/her work area.

Safety management includes the following tasks:

- setting safety objectives and risk acceptance criteria;
- organizing the safety responsibilities and work;

- motivating people for safety;
- identifying the hazards which must be addressed;
- assessing the risks associated with these hazards;
- implementing and following the results of risk assessment;
- monitoring and auditing the safety performance;
- planning for the potential emergencies.

Most of the above tasks are clearly the responsibility of the line management in the companies responsible for carrying out the petroleum activities. Some elements, however, are in many jurisdictions considered to be a regulatory responsibility, such as setting objectives and defining acceptance criteria, identifying hazards and assessing the associated risk, etc.

Regulatory authorities can choose different strategies in safety management, as evidenced by a shift that has occurred in Norwegian authorities' approach over the last several decades. Their original regulatory strategy was prescriptive, involving specific instructions often in much detail on how petroleum operations were to be conducted, and frequent inspections. More recently this has been shifted to a predominantly system-oriented and self-regulatory strategy, requiring operators to take their own responsibility for the complete range of management elements listed above, but involving also audits and verifications as well as dialog between operating companies and authorities.

The shift from a prescriptive strategy to a self-regulatory strategy has had the perceived effect of creating a less adverse and more cooperative relationship between authorities and petroleum operators in Norway. The safety performance in the Norwegian petroleum sector has improved in recent years (to 2007) and is favorable by relevant international comparisons.

To meet the goals set by the company management, it is necessary to set requirements applicable in the following areas:

- personnel's qualifications;
- technical safety;
- protection of the environment;
- management systems and procedures.

All personnel must be trained and qualified to perform their job, including the necessary qualifications for the safety aspects of their work.

The following principles may be applied in order to attain a high level of safety in petroleum operations:

- Safety has top priority as a criterion for decision making.
- Safety must as a minimum satisfy the regulatory requirements.
- Safety precautions are part of normal daily work.
- Safety is a line responsibility from the top management to each employee.
- Safety is integrated in all activities.
- Safety is verified in the line organization, and confirmed by an independent assessment where necessary.
- The company's operations are recorded in such a way that safety can be improved through experience.
- Any accident, damage, or loss is classified as a deviation from the norm and reported, so that corrective actions can be taken.

10.2 Organizing safety management

This section addresses several issues related to organizing the safety management of an operation or project.

10.2.1 Safety as a functional role in the organization

Industrial companies are organized in many different ways. The management of safety must be tailored to the overall organization and mission, and there is no "one and only" way in which to do this.

The responsibility for safety in an operation must always rest with the line management. In an engineering project, the project manager is responsible for safety. Many industrial organizations nevertheless find that they need a safety department, that is, a designated unit within the organization having functional responsibility for providing professional support and handling certain tasks in safety management. A safety department can be engaged in tasks on several levels. Following are examples:

- *safety management* as a process of establishing a consistent safety philosophy for the company, with safety goals, programs, etc.;
- *safety services*, including expertise in areas like risk assessment, emergency planning, industrial hygiene, etc.;
- *safety aspects in project planning*, including design of systems for fire fighting, gas detection, evacuation, etc.;
- *site safety*, including permit-to-work systems, gas measurements, safety introductions, emergency exercises, etc.

The development of an overall system in these areas in a company usually starts at the bottom, with site safety. As the company advances, the need emerges for a systematic approach to safety in projects, using risk assessment tools to make engineering decisions. Finally, the need to organize the total approach to safety in a structured safety management system is recognized.

10.2.2 Improving safety management over time

As an industry matures globally or within a country, it may increase the level of sophistication with which it manages safety. A common development pattern is as follows:

Phase 1: Industrial activity evolves without any systematic risk analysis, leading to unawareness of major hazards and a shortfall of measures to control them.
Phase 2: A major accident takes place, which shakes up the industry and leads to regulatory action.
Phase 3: Risk analysis is introduced, with strong emphasis on risk quantification and comparison with "acceptable risk targets." This leads to a major step forward in the way major hazards are considered in design.
Phase 4: After a period of risk analysis application, it can be seen that, although risk analysis brought things forward, it is not the only thing needed. Authorities perceive that the industry is playing a "quantification game" rather than concentrating on the real safety issues.
Phase 5: Safety management is introduced as a framework, including risk analysis. Authorities' focus shifts from reading risk analysis reports to reviewing how the results are being

used in decision-making processes, the steps taken to implement the recommendations, and ensuring that the assumptions are valid.

10.2.3 Learning from experience (feedback loop)

Accidents are usually unexpected, and happen in spite of measures that were meant to provide safety. The challenges of managing safety evolve over time. Successful safety management for the long term therefore requires that the organization systematically learns from experience, enabling it to reflect experiences gained in future decisions and priorities, including changes to its safety plans as may be required. Figure 10.1 illustrates this learning process as an "improvement loop."

Learning processes in this area should not be confined to any particular company. Authorities have an important role in recording and conveying relevant information and experiences, including unwanted events, across the boundaries of individual organizations engaged in petroleum activities. This is an area where firms which are otherwise competitors may exchange information and views in order to improve safety in the entire industry.

10.2.4 Safety management of contractors

The petroleum industry uses contractor services extensively for engineering, construction, and operations. The company's safety management system cannot work adequately unless a similar level of efforts and organization is required from contractors.

Each contractor must therefore be required to adopt a safety management system which is compatible with the company's system. This must be initiated at the stage of contractors tendering for the services, by laying down the requirements for safety management as part of the contract. This requirement must be followed up at management level.

10.2.5 Safety planning and follow-up in a petroleum development project

In the engineering phase of a project, any activities related to safety management must be conducted in a systematic manner. A *safety plan* (safety program) can provide for this. The purpose of the plan is to ensure the following purposes:

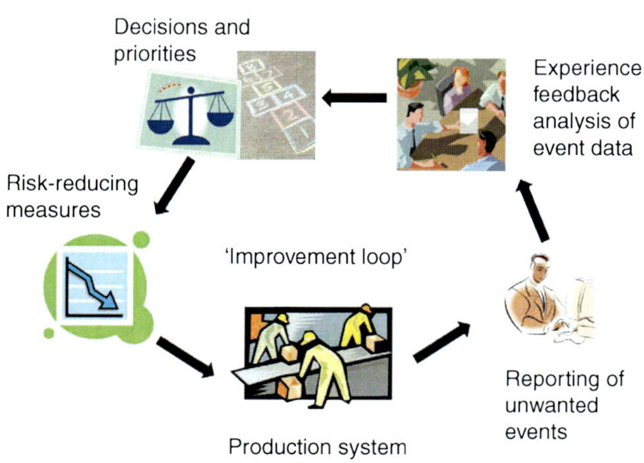

Figure 10.1: Improvement loop for safety management.

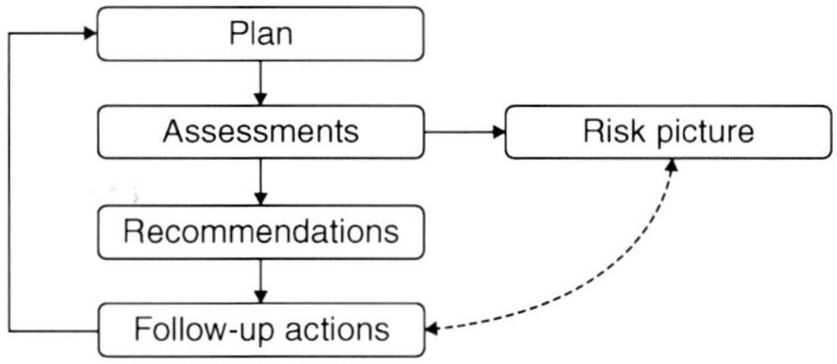

Figure 10.2: The role of the safety plan.

- management control over safety aspects in the design process;
- ensuring that safety aspects are accounted for in a timely way in the design process;
- consistency in decisions to be taken;
- traceability and documentability of the safety aspects in the design basis.

The plan should include the following elements:

- description of roles and responsibilities within the project;
- identification of the planned assessments (hazard and operability reviews, risk analysis, and emergency preparedness);
- description of methods and responsibilities for updating the plan (including the list of planned activities);
- system description for follow-up of findings from the assessments.

Figure 10.2 illustrates a work process for assessing and following up risks related to a development project under the guidance of a safety plan.

Large petroleum development projects usually require a number of assessments of risk aspects and emergency preparedness. Each of these normally results in several recommendations. The assessments require certain assumptions to be made as a basis for analysis. An important task in a development project is to follow up all recommendations systematically and to ensure the continued validity of the assumptions made for the assessments.

For following up on the assessments, the safety plan needs to clarify responsibilities for following up individual actions. A systematic procedure for tracking the status of proposed actions and recommendations is also required.

In practice, a database is often helpful for tracking all recommendations and to allow that these can be checked out against the status in the project. The implementation of actions and recommendations resulting from a safety study or risk assessment may alter the risk picture of a facility. The follow-up system should therefore allow reconsidering the risk picture when the recommendations are closed out.

10.3 Safety management at the stages of a development project

Safety must be incorporated as a key criterion for the planning and design of a petroleum project. The technical risk for each area of the installation must be identified and compared with the acceptable level specified in internal documents. The overall technical risk must be identified and evaluated as acceptable before main decisions are taken.

This section describes the safety management actions and systems during the different phases of a petroleum development project. The progression of a development project is described in Chapter 8. The section is divided in subsections that cover the relevant phases, as follows:

- feasibility phase;
- concept selection phase;
- front end engineering (pre-engineering) phase (FEED);
- detailed engineering phase;
- construction and installation phase.

Each section starts with a brief description of the relevant phase, before addressing the safety management aspects that may be important for that phase.

10.3.1 Feasibility phase

In the feasibility phase of project development, the information available on design is tentative, since no development concept has yet been selected. There is not much scope for performing extensive safety studies in order to identify a feasible concept. Instead, assessments may be performed on the basis of the generic characteristics of the development concepts, with assessments and possibly ranking of advantages and disadvantages in terms of safety. The factors that may be considered at this stage may include:

- field layout;
- amount of processing facilities (complexity as well as hydrocarbon inventories);
- separation distance between processing facilities and living quarters;
- number of wells;
- for an offshore development, whether wells will be subsea wells or drilled from a fixed installation;
- choice between export pipeline or other offtake system;
- need for simultaneous operations (drilling and production) and required restrictions;
- number of export/import pipelines;
- the need for flow control by chemical injection or flow-line heating;
- offshore: choice of support structure for the installation (floating and fixed);
- size of accommodation quarters.

10.3.2 Concept selection phase

At this stage in the project, one will aim to identify the most optimal field development concept and identify possible safety-related showstoppers for any of the concepts considered. Any main features of each concept that are particularly important for safety need to be identified and assessed.

Such features can be identified by assessing the differences between the various alternative concepts in terms of risk and safety. This can start with the identification of known or expected hazards. The relevance of each such hazard can be assessed for each development concept. Such potential hazards may include:

- fire or explosion caused by a leak from the process;
- blow-out;

- release of toxic gases or liquids (e.g., H_2S in crude or gas);
- leak from a riser or pipeline (and subsequent fire and/or explosion);
- offshore: ship collision with the petroleum installation;
- helicopter accident;
- explosion in crude storage tanks;
- dropped object (particularly on subsea facilities);
- fire in utility systems, including electric, diesel, chemicals, or helifuel systems;
- fire in living quarters;
- earthquake;
- extreme weather conditions, landslides, icebergs, and extreme icing;
- fatigue or wear of the support structure of the installation.

Other risk factors than those listed above may also be relevant, often dependent on the location of the field development. Local conditions related to the surface environment or reservoirs may affect the risk level of otherwise well-known hazards.

Both in the feasibility phase and when performing concept studies, experienced personnel will be needed to carry out safety studies, because only coarse information is available that requires to be interpreted. The person(s) involved must have sufficient understanding of the type of facilities and operations considered and about the risks that may be involved.

10.3.3 Front end engineering phase (pre-engineering)

In the FEED phase, the concept considered will be better defined than in the previous phases. Overall dimensions and layouts are developed, as well as the number of wells to be drilled and the main process parameters.

Normally in a project, the conceptual design will be sufficiently detailed to perform a quantitative risk assessment of the concept. In Norway, the term concept safety evaluation (CSE) has been in use since 1981.

Depending on the concept chosen, it may be useful to perform further safety studies in addition to the CSE. If a new or special design is chosen, a safety study of that specific aspect may be needed in order to assess the impact of the design on safety.

The safety management work in the FEED phase must form the basis for planning the detailed engineering phase, and should identify any impact on budget and schedule. The safety work must therefore focus on identifying:

- any features that could result in an unacceptably high level of risk;
- any measures needed to correct features that represent high risk;
- any high-risk features that need to be evaluated further;
- the risk factors that may be controlled by proper design or operations.

10.3.4 Detailed engineering phase

In the detailed engineering phase, the selected concept will be developed further into a complete and detailed design. This includes all documentation for constructing and installing the facilities. Also documentation for the operation phase will need to be developed as part of the detailed engineering phase.

Safety management efforts in the detailed engineering phase may concern two main purposes:

(1) *Definition of design requirements for the various design disciplines*: In particular, this includes dimensioning loads and requirements to safety systems and emergency preparedness (design accidental loads (DAL)), and may include:
   - fire and explosion design loads (basis for design of fire protection and fire and blast walls);
   - ship collision loads;
   - requirements for evacuation means, including evacuation capacity;
   - requirements for fire and gas detection systems;
   - development of a fire water philosophy;
   - development of an emergency shutdown philosophy.

(2) *Verifications of the design work*: In addition to verification of the process design, recommendations from risk assessments must be followed up. Tasks may include, for example:
   - performing hazard and operability studies (HAZOPs) on the design of process and utility systems;
   - monitoring the design in areas that are associated with a higher risk, for example, wellhead areas;
   - monitoring the consistency of design development in relation to recommendations and assumptions from risk assessments;
   - identifying any developments or deviations in the design that may affect the risk level of the new facilities.

10.3.5 Construction and installation phase

In the construction and installation phase, most of the engineering work has been completed, and focus is now on building and installing facilities in the field. The safety issues that may need attention in this phase include:

- workplace safety on construction and fabrication sites;
- safety critical operations such as lifting and marine operations;
- safety and coordination of simultaneous operations, like drilling and installation;
- preparation for production operations including training.

There is a need to ensure that the installations are being built as intended in the design, and to identify and resolve any misunderstandings in an early stage. Training of operations personnel and preparations for start-up may also be required. Often some apparently minor issues arise in a project, requiring a solution on short notice because of the time pressure in the project. An important challenge is to prevent such situations from resulting in unacceptable "shortcuts," that is, bypassing regular procedures and quality safeguards.

Construction and installation works are usually performed by a number of contractors that often employ large resources in terms of both manpower and materials. Specialized subcontractors are often used to perform parts of the installation work like: laying subsea pipe, lifting of modules, installing subsea equipment, etc. These contractors need to be coordinated to ensure that they do not interfere with each others' operations.

Special attention should be paid to simultaneous activities. Two or more activities that are performed in the same area at the same time may interfere with each other. Especially communication lines, position reference systems (for vessels), and anchors can be vulnerable for interference phenomena, and access to work sites may be restricted. A *HAZOP*, if set up properly with involvement of all relevant parties, is a good tool for identifying any potential for interference and allowing this to be resolved.

Even though the types of activities may differ from the engineering phase, the management systems (safety plan, responsibilities, etc.) should remain the same during the construction and installation phase.

10.4 Risk assessment

Risk is associated with the possibility of an accident or other unfavorable deviation from a norm. Risk is regarded as the logical opposite of safety, and is therefore used as a measure of safety.

Risk can be defined as the product of probability and consequence of a potential event:

$$\text{Risk} = \text{Probability} \times \text{Consequence}$$

Probability is the likelihood of a potential event. It is expressed as a value in the range of 0–1: a probability of 0.00 is a certainty that the event will not happen, a probability of 0.01 represents a small possibility that it may happen, and a probability of 1.00 is a certainty that it will happen.

Consequence is the severity of negative impact if the accident does happen.

The simple formula shown above can be used as a basis for *quantitative* risk assessment. This aims at expressing risk in a numerical term which can be compared against a criterion for what is considered acceptable or not. A flow diagram for quantitative risk assessment is shown in Figure 10.3, indicating the main steps involved in performing the assessment.

In the sections below, the frequency and probability assessment and consequence analysis are described. Hazard identification is addressed only briefly.

Figure 10.3 also may be valid for *qualitative* assessments. In that case, the different activities are not based on quantified models and calculations, but rather on the judgment of the personnel involved in the assessment.

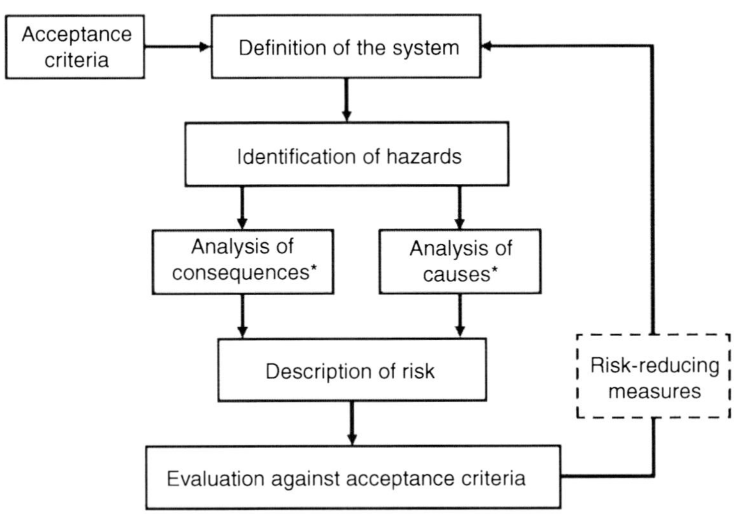

Figure 10.3: Elements of a risk analysis. (*) Qualitative or quantitative analysis.

10.4.1 Acceptance criteria

Acceptance criteria are verbal statements or numerical parameters which express the level of risk or reliability deemed acceptable or desired in particular situations. Acceptance criteria may be determined as regulatory requirements or company policy.

10.4.2 Methodology for quantitative risk analysis (QRA)

Quantitative risk assessments are utilized to estimate the risk from potential undesirable events. The risks are quantified in terms of the probability and consequences of undesirable events. By comparison of the quantified risk level with suitable risk criteria, the analysis can be used to help deciding whether the risk level is satisfactory, or whether risk-reducing measures will be necessary.

10.4.3 Hazard identification

The basis for any risk assessment, whether it may be qualitative or quantitative, is the *hazard identification*. This analysis should be performed in a systematic way, to ensure that all hazards are being identified. This can be accomplished either by using one of the available techniques or by developing a purpose-built approach. Many of the techniques can be classified as being either deductive in nature or inductive.

A deductive technique is one where a "top event" is formulated and where the causes for that top event are being identified. Setting up a *fault tree* is an example of a deductive technique. Fault trees (Figure 10.4) graphically represent the interaction of failures and other events within a system. Basic events at the bottom of the fault tree are linked via logic symbols "and" or "or" (known as gates) to one or more top events. These top events represent

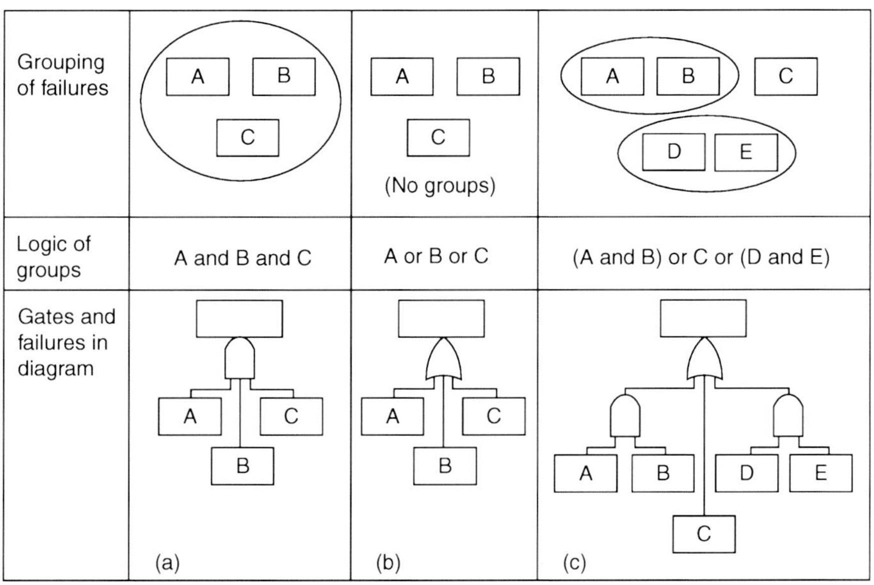

Figure 10.4: Gates in a fault tree.

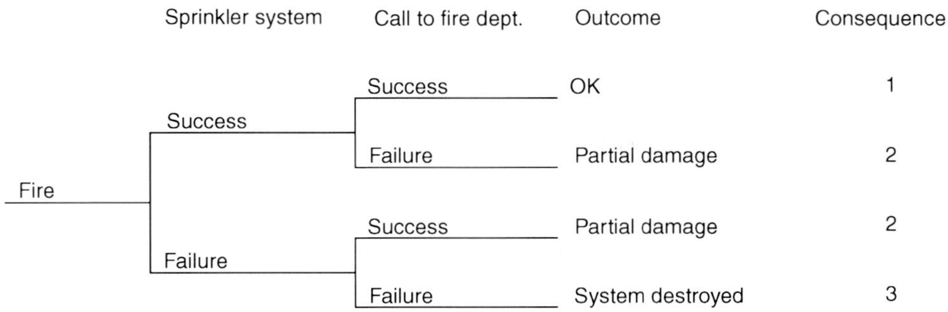

Figure 10.5: Simple event tree.

identified hazards or system failure modes for which predicted reliability or availability data are required.

An inductive technique starts with formulating a deviation or a fault and then attempts to identify the possible consequences of that deviation or fault. The process of developing an *event tree* is an example of an inductive technique; see Figure 10.5. In risk analysis, models are used to describe the consequences of leaks or other accidents. Models and calculation techniques exist for the different phenomena that may occur after a leak: outflow, gas dispersion, fire, explosion, etc. These phenomena together make the accident scenario.

The risk analyst has to select the models which he considers suitable for her or his application. When doing this, the analyst must recognize that the outcome of the analysis depends on the model chosen and on the way it is used. The appropriate model must be selected based on the knowledge of the available models and their limitations.

10.4.4 Frequency and probability calculations

In order to carry out a QRA, it is necessary to identify the events which could lead to accidents, the initiating events, and their probability (frequencies).

10.4.4.1 Initiating event frequency

Initiating events are transients which trigger a sequence of events that may end in some sort of an accident. A systematic hazard identification analysis can be used to identify a complete set of initiating events. Typical examples of initiating events may be:

- a gas leak from a process segment;
- a rupture of a liquid storage tank;
- a well kick causing drilling mud to be ejected;
- a power black-out;
- an object being dropped from a crane;
- a tanker vessel coming adrift toward offshore installations.

An initiating event places a "demand" on the built-in safety barriers and emergency systems of the installation, and within the local community if third-party (public) risk is concerned. Estimates can be made for how often the demand is likely to occur, that is, the initiating event frequency. Historical events have been summarized in databases which can be referred to in order to document the frequency estimates; reference is, for example, made to the OREDA database (Offshore Reliability Database, managed by Sintef, Trondheim).

10.4.4.2 Barrier unavailability

Risk analysis is based on the concept of an accident being the result of some unwanted release of energy or some toxic material. To prevent this from happening, the energy is controlled through a set of barriers. Hence, risk analysis concentrates on the integrity of these barriers and aims to quantify their probability of failure.

A *barrier* may be a physical installation as well as various features and operating procedures (Figure 10.6). For example, we may provide a "barrier" by installing an offshore platform such that the living quarters are upwind most of the time, to reduce the risk of gas or smoke entering the living quarter. The wind direction is thus a barrier.

10.4.4.3 Event sequence frequency

After having identified the initiating event frequencies and the branching probabilities in the event tree, the next step is to estimate the frequency of each consequence.

10.4.4.4 Summary of risk results

Most risk analyses will be made up by a comprehensive set of event trees, covering different initiating events for each type of hazard. It is important that the final consequences are categorized in a coherent way, in order to allow a summary of the risk results to be made up.

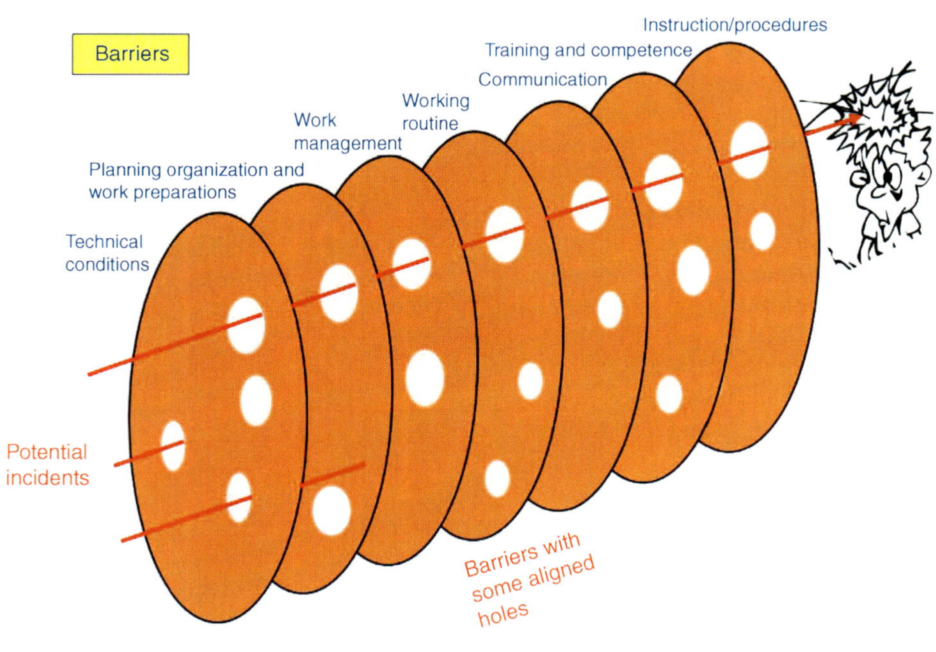

Figure 10.6: Barriers with some aligned holes.

This is done by simply adding the frequencies of all consequences belonging to the same category. The result is a table giving the pairs of frequencies and associated consequences for the whole installation, offshore plant, etc.

10.4.5 Presentation of the risk assessment

The method best suited for communicating the assessed risk varies with the application. In some circumstances, it is convenient to use a single number—one that is often used is the average fatal risk to the workers or the public. However, a single number is usually inappropriate as it does not reveal who is at risk or whether the risk is due to a few multiple fatality incidents or several single fatalities. For this reason, graphical presentations of risk are often preferred.

A risk matrix, shown in Figure 10.7, is a common form of graphical presentation of risk. Individual sources of risk can be plotted against the two dimensions of frequency (which is closely linked to probability) and consequence, providing indications of whether the risk is high or low.

Depending on the end use, either individual or societal risk assessments are appropriate. The normal method of presenting individual risk information for onshore facilities is the well-known plot of risk contours superimposed over a map of the area. This is relatively well understood by industry, the government, and many in the general public.

Risk contours provide an incomplete indication of how an accident would affect nearby residents, since this also depends on the population distribution. Society often views multiple fatality incidents more severely than an equal number of single fatality incidents, and regulatory authorities and companies assent to this view. The usual method of displaying societal risk is by an *f–N* plot, which shows, on a log/log scale, the estimated frequency (*f*) of incidents causing *N* or more fatalities. Three regions can be defined on a plot—one in which the

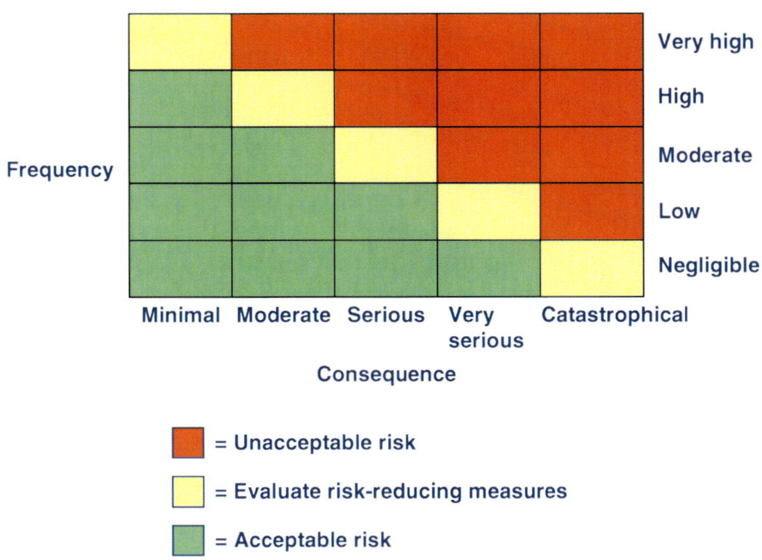

Figure 10.7: Risk matrix.

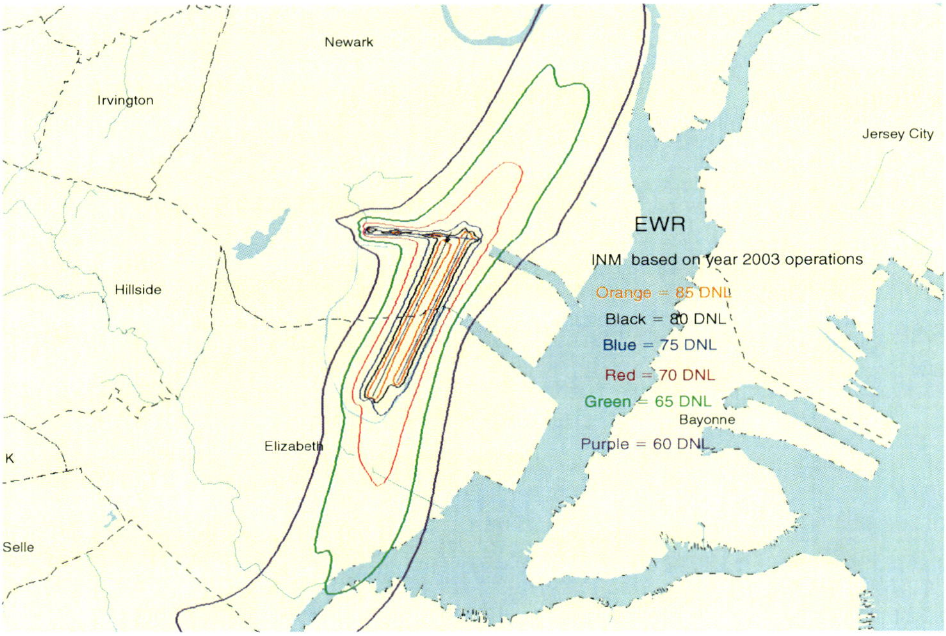

Figure 10.8: A noise contour is a line on a map that represents equal levels of noise exposure. Illustration: Boeing.

risk is considered low, another in which the risk is considered unacceptable and improvements are mandatory, and an intermediate one in which the acceptability is re-assessed taking into account the cost of improvements.

This can also be shown with a risk matrix; see Figure 10.7. The risk matrix that is shown in Figure 10.7 represents the outcome of a qualitative analysis.

Figure 10.8 shows the use of noise contours for an airport, which is a variation on the use of risk contours as may be done in the petroleum industry. In the case shown in Figure 10.8, the Integrated Noise Model (INM) is used to calculate annual noise contours. The number of annual flights, aircraft types, and flight tracks are critical inputs to the contour model. The attached map presents the modeled dynamic noise level (DNL) contours for 2003 operations shown in 5-dB increments for DNL values of 60, 65, 70, and 75 dBA for Newark Airport in the state of NJ, USA. (*Source*: Boeing, www.boeing.com.)

10.4.6 Cost–benefit considerations

Quantitative techniques can be used for cost–benefit analysis. A cost–benefit analysis can be performed as a basis for deciding whether or not a certain measure is worth implementing. It is best suited for economic decision processes, where the cost of an investment must be weighted against the expected, but uncertain, revenues from the investment. The cost–benefit analysis results in a comparison of dollars against dollars. It can also be applied to judge the cost effectiveness of safety investments; however, one should realize that cost–benefit analysis normally would not be appropriate when weighing lives against economic values.

Environmental management

The petroleum industry has a potential for inflicting serious damage on the environment, and a history of doing so in several parts of the world. Damage may come from pollution (not least from oil itself if released to the environment), or from excessive impacts of installations and operations. Now, the petroleum industry must work under the presumption that authorities and general society have very low tolerances of such damage. Environmental damage can also inflict large financial as well as reputational costs on a company, as evidenced by such cases as the Exxon Valdez accident in Alaska in 1989.

Environmental management concerns the work processes and designs required to avoid unacceptable environmental impacts from industrial activities.

There is some overlap between the issues of environmental management and safety management as exposed in the previous chapter. Safety management concerns the avoidance of accidents, and accidents in petroleum operations would in many cases result in environmental damage as well as harm to people and assets. Safety management thus is an indispensable contributor to environmental management. This chapter will review mainly those aspects of environmental management that are additional to safety issues as reviewed in Chapter 10.

The chapter aims to outline some useful principles and guidelines for environmental management. The following sections summarize the main principles of the environmental management, explain the main types of discharges and emissions, and present the design principles for reduced environmental impact. Environmental risk and acceptance criteria for the environment linked to oil spill and contingency planning are also covered in this chapter.

The chapter draws extensively on practices from the Norwegian petroleum sector. This sector has been subjected to stringent environmental standards in harsh, mostly offshore environments over a long period of time. It is believed that some of those practices can have useful applications also elsewhere, even if natural conditions are much different.

11.1 The need for sustainable petroleum operations

The most basic mission of the oil industry is to extract petroleum from the ground, and then to make petroleum useful and available to customers. In this simple context, measures to protect the environment may have been viewed as an imposed burden, involving regulations, added cost, and liability. This viewpoint must now be considered outdated, and would leave firms ill-equipped to deal with the issue in a strategic manner. Petroleum firms have been challenged to redefine their mission in order to conduct their operations in harmony with a wide range of important societal needs. One modern approach to this challenge is the notion of the *sustainable enterprise*.

A sustainable enterprise is a firm whose activities are sustainable in terms of economics, social effects, and the environment. *Sustainable* activities require that value is created, not destroyed, in the long term. This requirement must be met on each of the three levels: the economic, social, and environmental levels. Economic sustainability requires that the firm is profitable as a result of using resources effectively. Social sustainability requires that the firm contributes positively to the well-being of people and societies affected by its operations. Environmental sustainability requires that no serious damage is inflicted.

Conducting petroleum operations without harming the environment requires two things:

- understanding the natural environment and its sensitivities to man-made impacts;
- an effective framework for environmental management, based on public regulations as well as managerial systems within the firms.

Many large petroleum firms have established corporate policies for environmental sustainability. As an example, Statoil has set a policy of zero harm to the environment, which it defines as follows:

Conserving biodiversity:

- no habitat destruction;
- no introduction of foreign species;
- no effects on population levels of species.

Limiting emissions and discharges:

- emissions and discharges to be below the critical level of relevant ecosystems.

Limiting land use:

- restore and clean used areas when activity is completed;
- conserve landscape and cultural heritage.

Petroleum operations are often based on constructing large installations which will operate for many years. The design solutions selected initially will influence operations far into the future. The environmental management of a development project needs to be initiated at the earliest stage, when the possibility of influencing the concept and the technology selection is still high. Consequently, environmental protection has to be considered from day 1 of developing petroleum resources.

11.2 Regulating environmental management: the Norwegian approach

Legislation and the regulatory approach to environmental protection differ between countries, due to their different legislative systems, histories, and natural and industrial conditions. The legislation and regulations concerning environmental protection applicable to petroleum exploration and production in Norway are presented here. This may provide a relevant reference for the most important environmental regulatory issues elsewhere.

The applicable legislation and regulations for Norway are available in Norwegian and English on the website of the Petroleum Safety Authority (www.ptil.no).

11.2.1 The regulatory approach

The Norwegian safety regulations for petroleum activities state that it is the operator's responsibility to ensure an acceptable level of safety within all operations and activities. Operators are obliged to define risk acceptance criteria and to perform necessary risk assessments to ensure an acceptable safety level. The term "safety" in this context includes the safety of personnel, the environment, assets, and production. The safety regulations are applicable to all petroleum activities, regardless of location, type of installation, or operational conditions.

The safety regulations contain functional rather than prescriptive requirements. Functional requirements specify what the regulator wants to achieve, whereas prescriptive requirements would specify the solution required. The Norwegian approach to these regulations has shifted from prescriptive to functional requirements, thereby placing responsibility for safety management more clearly with the operators.

Norwegian environmental regulations make operators responsible for securing an acceptable level of safety for the environment. This includes defining environmental goals, assessing environmental risks, selecting strategies for reaching the goals, selecting environmentally friendly technologies, minimizing regular discharges and emissions, and establishing the necessary management system to meet the environmental requirements.

11.2.2 Legislation and regulations

The most important legislation for the environmental aspects of the Norwegian "upstream" petroleum sector is the following:

- The Petroleum Act;
- The Pollution and Waste Act;
- The Product Control Act;
- The CO_2 Tax Act.

In addition, projects which are fully or partly located onshore are affected by the *Planning and Building Act.*

The Acts are passed by Parliament, and determine the main principles, responsibilities, and requirements. *Regulations* are issued by the Cabinet as authorized by an Act, and specify responsibilities and requirements in more detail. Regulations that have been issued on the basis of the Acts listed above include requirements for the following:

- environmental management system;
- environmental impact assessment (EIA);
- acceptance criteria for environmental risk;
- environmental risk assessment;
- emergency preparedness analysis;
- oil spill contingency plan;
- application for discharge permit;
- waste management plan;
- monitoring program.

Section 1-2 of the Petroleum Act, regarding resource management, states that "Resource management of petroleum resources shall be carried out in a long-term perspective for the benefit of the Norwegian society as a whole. In this regard the resource management shall provide revenues to the country and shall contribute to ensuring welfare, employment and an improved environment, as well as to the strengthening of Norwegian trade and industry and industrial development, and at the same time take due regard to regional and local policy considerations and other activities."

The Pollution Control Act contains rules applicable to the offshore exploration and exploitation of petroleum. Provisions concerning permissions and regulations do, however, only apply to the activities which regularly lead to pollution, as risk for acute spills is regulated by the Norwegian Petroleum Directorate (www.npd.no). The Pollution Control Act expresses the general duty to protect the natural environment from pollution and waste problems. Efforts shall be made to prevent pollution from occurring, limit existing pollution, and avoid waste problems. Unless a discharge permit is given by the Climate and Pollution Agency (www.klif.no), on application, any pollution is prohibited.

In this context, pollution is defined as the release to air, water, or ground of solid matter, fluid or gas, noise and vibrations, light, radiation, and temperature modifications, which cause damage to the environment.

11.2.3 Approval processes

The regulatory system typically includes three important application and approval processes:

- application for petroleum production;
- application for discharge permit;
- application for consent to drill.

Approval for petroleum production in Norway is given as an approval of the Plan for Development and Operation (PDO), including the EIA report. The approval is granted by either the Parliament or the Ministry for Petroleum and Energy, depending on the overall development cost.

A discharge permit is given by the Climate and Pollution Agency. The Petroleum Safety Authority approves the consent to drill. All processes include public consultations, and conditions may be linked to the approval.

In addition, an important means for regulatory control is the process preceding the opening of new areas for petroleum operations. In this extensive process, in which the EIA is a central process, the authorities may provide for protection of particularly sensitive areas.

11.2.4 Environmental impact assessment

An EIA is a comprehensive analysis of a proposed action in terms of its expected impacts on the environment and on society. An EIA should also propose mitigating measures that can avoid, reduce, or compensate for negative impacts, as well as measures to enhance positive impacts of the proposed action.

EIAs, and the regulations applicable to them, provide a formalized framework for communicating such impacts to decision makers (including companies and authorities), stakeholders, affected communities, and the general public. It should provide for the participation of affected parties and the general public in decisions leading to petroleum developments, thereby improving the quality of decisions and reducing the risk of unanticipated negative consequences.

EIA is a well-established concept in many countries, and may be applicable to a range of other projects besides petroleum developments. For Norwegian petroleum developments, an EIA is normally needed for project approval and is a mandatory part of the PDO.

EIA is required at three stages of Norwegian petroleum developments:

- preparation of EIA before introducing petroleum activities in a new region;
- preparation of a regional EIA (REIA);
- preparation of a field-specific EIA (FEIA).

EIA procedures in Norway require a study program to be prepared ahead of the EIA itself. The study program outlines the project and the content of the EIA. The study program is submitted to public consultation and approval by the competent authority. This serves to ensure that the EIA itself covers all relevant issues in terms of how the project may affect the environment and the public.

In many cases, various study reports on specific issues are prepared in support of the EIA. The EIA main report includes a description of the project with alternative concepts, areas of influence, environmental and social impacts, mitigation measures, recommendations, and a management plan for following up and monitoring the impacts of the project.

The EIA and supporting documents are prepared by the project operator, usually employing consultants. The EIA is submitted to public consultation and approval by the competent authority in conjunction with granting required approvals for the project itself.

11.2.5 Regulation on the use of chemicals

Companies must apply for discharge permits from the Climate and Pollution Agency for the discharge of oil or chemicals to sea. The Authority grants discharge permits pursuant to the provisions of the Pollution Act. Under the Act, the operating companies themselves are responsible for and obliged to establish the necessary contingency planning measures to counteract acute pollution. Municipal and national emergency response plans are also in place (Ministry of Petroleum and Energy, 2006, p. 61).

"Chemicals" is used here as a generic term for all additives and auxiliary products used in drilling and other operations. The main rule is that no environmentally hazardous substances may be discharged, regardless of whether the substance is an additive or occurs naturally. A majority of the chemicals used in the Norwegian petroleum industry consist of substances that are believed to have little or no impact on the environment. They have then been classified as "green" or "yellow" chemicals by the Climate and Pollution Agency. Many such chemicals are substances that occur naturally in seawater. The remaining chemicals have impacts which are environmentally hazardous or which have not been adequately documented. Most chemical discharge is associated with drilling operations.

The environmental impact of chemical discharges may be considered in terms of the following factors:

- the type of compound and the severity of its environmental impact;
- the quantities used and discharged;
- where and under what conditions they are released relative to the conditions in the recipient environment.

Discharged chemicals may cause a local contamination effect. Research has shown that chemicals tend to be diluted in the water column so that they do not cause a significant acute environmental impact beyond the immediate locality of the release. Certain chemicals, however, may have highly dangerous environmental consequences, including hormonal disruption and bioaccumulation.

There is still uncertainty about any long-term effects of discharges into the sea, but considerable research is being devoted to the issue (Ministry of Petroleum and Energy, 2005).

11.2.6 Waste management

The Pollution Control Act contains requirement for waste management. The primary strategy is to avoid the generation of waste and reduce discharges. Operators are required by law to implement a suitable waste management system.

11.3 Environmental conditions and natural resources

The environmental challenges facing the petroleum industry include the requirements stipulated by the authorities as well as the precautions necessitated by the environment itself. This section will outline some characteristics of the environment that may influence the development and operation of petroleum projects.

There is a need for mapping the environmental conditions of an area preferably before starting exploration activity. A good understanding of environmental constraints facilitates the selection of suitable development strategies, and thus to avoid expensive changes or repairs later in the project.

11.3.1 Environmental sensitivity and sensitivity mapping

Different ecosystems, and plant and animal species have different sensitivities. The term "sensitivity" in the ecological context is used to characterize the tolerance of organisms to negative impacts such as pollution.

Environmental sensitivity mapping is the process of summarising ecological information from literature or field studies in a report or a map. An example is shown in Figure 11.1. Information on the biological resource distribution in the area and the vulnerability to the negative impact in question is presented. For The West Coast of Ireland, see Wood et al., 1996.

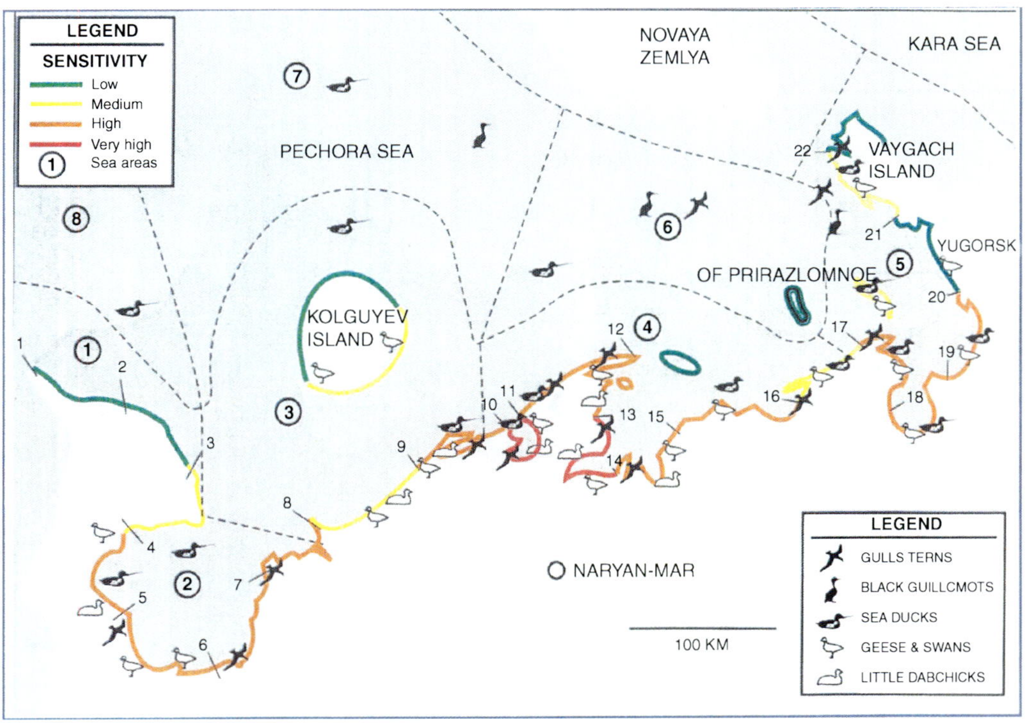

Figure 11.1: Distribution and sensitivity of fish in the Pechora Sea in ice-free water in the summer (Gudmestad et al., 1999).

Furthermore, Environment & Resource Technology Ltd, 1997, presents the seasonal variations in species occurrence and their susceptibility to pollutants for The Faroe Islands.

The functioning of ecosystems is influenced by physical, chemical, as well as biological factors, and these ecological factors are regarded as essential for understanding the various processes in ecosystems.

Seasonal and geographical variability in the physical–chemical environment affects all marine life, and consequently living organisms exhibit varying capabilities for ecological adaptations to changing conditions of existence. There are seasonal and geographical fluctuations in distributions and qualitative and quantitative characteristics of the marine communities. Natural fluctuation is recognized as an inherent feature of the marine populations, and undisturbed marine communities are regarded as being stable or in a state of equilibrium.

11.3.2 Environmental baseline survey

Before petroleum operations begin in a pristine area, the existing environmental situation needs to be documented. This is done with an environmental baseline survey. Baseline results should preferably be acknowledged by the authorities prior to any physical work on site.

Existing baseline data of the environment and society may be assessed in terms of quality. If available data are insufficient for project development or for documenting the impact of the planned activities, survey(s) must be undertaken. The survey program must be in accordance with applicable regulations and adjusted to the size, properties, and location of the project, that is, an exploration well will have a smaller survey than a development of an area.

A similar requirement applies for a company planning to acquire existing facilities and operations. In this case, particular emphasis must be given to the current pollution situation in the influence area. The study program should be tailored to reveal possible legacies that might become the company's responsibility.

In an environmental baseline survey, the impacts of the proposed action on the listed resources within the action area are assessed. The influence on the survival and recovery of the species or the overall value of the designated critical habitat can then be analyzed.

11.4 Environmental management systems and tools

Operators of petroleum ventures are under legal requirements (in Norway and other jurisdictions) to implement an appropriate environmental management system in their organizations. Many companies apply such policies to their global operations, which can be more stringent than national regulations. To be effective, the organizations' efforts in environmental protection need to be conducted within a structured management system.

11.4.1 Environmental management standards and guidelines

National authorities may set requirements for management systems for compliance with statutory requirements in relation to safety and environmental protection, as reviewed elsewhere in this chapter. In addition, international organizations have developed helpful guidelines and standards for development and implementation of environmental management systems. Such guidelines have been provided jointly by the International Association of Oil & Gas Producers (OGP), UNEP Industry and Environment (E&P Forum, 1994), and the International Standardization Organization (ISO, 1996), respectively.

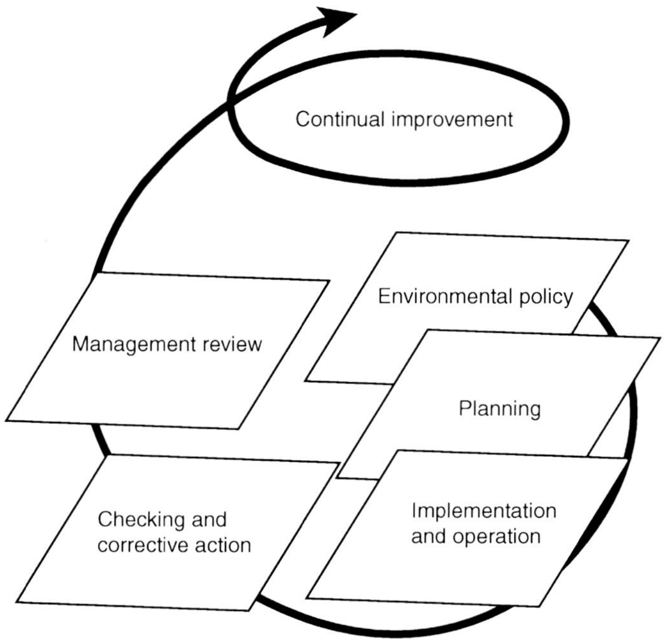

Figure 11.2: Environmental management system model from the ISO 14001 standard.

These documents present various system models with slightly different content or view, but aiming toward the same goals of securing good environmental performance.

The international standard ISO 14001:1996 (Environmental management systems-Specifications with guidance for use) specifies the requirements of an environmental management system. The overall aim of this international standard is: "to support environmental protection and prevention of pollution in balance with socio-economic needs." The environmental management system model from ISO 14001 is outlined in Figure 11.2.

ISO 14001 emphasizes two important principles for successful environmental management:

- top management commitment to environmental performance;
- continuous improvement in effective performance.

Proper environmental care within the petroleum business is only possible with a strong company commitment to environmental protection. Environmental management needs to be an integral part of overall management activity in an organization. Senior management should therefore communicate the policy and objectives developed thereof clearly, as well as provide for the necessary resources to develop and maintain the system and support project teams and personnel during planning and operations.

11.4.2 Management tools

In addition to the elements described in the ISO standard and E&P Forum/UNEP guidelines regarding major components in the company management system, several tools are used to facilitate environmental management at the operational level, like the Environmental Program and Environmental Accounting System.

An *Environmental Program* is a useful tool in petroleum development projects and operations. This program may consist of three parts:

(1) the management principles applicable to the specific project, activity, or operation;
(2) objectives, requirements, acceptance criteria, and strategy;
(3) activity lists to secure that the goals are fulfilled.

An *Environmental Accounting System* is a tool for effective recording and reporting of various data for usage and discharges. The system provides the necessary updated environmental information for analyzing and reporting whenever needed.

Monitoring is a key element in a system of internal control (self-regulation). It is also needed for the purpose of compliance with regulatory and corporate requirements. It is a tool for measuring the effectiveness of operational procedures, and for identifying unexpected changes or unwanted impact of developments and operations.

The Norwegian Pollution Control Authorities (SFT) has issued guidelines for the surveying on the Norwegian Continental Shelf: "Environmental monitoring of petroleum activities on the Norwegian shelf: guidelines."

To facilitate a rapid and effective compilation of environmental data that are needed for a risk assessment or EIA execution, some oil companies at the Norwegian Shelf joined efforts and established the *Marine Resource Database* (MRDB). The MRDB contains updated information about resource distribution and area use, protective status, etc., in the Norwegian water and coastline. Maps and figures can be produced from the database, to view the environmental characteristics that are baseline input for the risk and impact assessment work.

A *waste management plan* is a useful tool for providing proper waste management within the company or a project. Waste management should aim to avoid waste generation. When waste is generated, focus should be put on re-using, recycling, or recovering the waste.

11.5 Operational discharges and impact evaluation

The environmental impacts of the offshore petroleum industry can be grouped into the following, as illustrated in Figure 11.3:

- discharges to sea;
- emissions to air;
- production of waste.

This section provides an overview of emissions and discharges from offshore petroleum activities based on information from Norway. Details are omitted, as the purpose here is to illustrate the link between type and amount of emissions and discharges, the potential effects on the environment, and alternative pollution reduction measures to be evaluated.

11.5.1 Discharges to sea

The main groups of discharges to sea from the offshore petroleum activities are:

- oil in water (originating from produced water, displacement water, and drainage water);
- accidental spills;

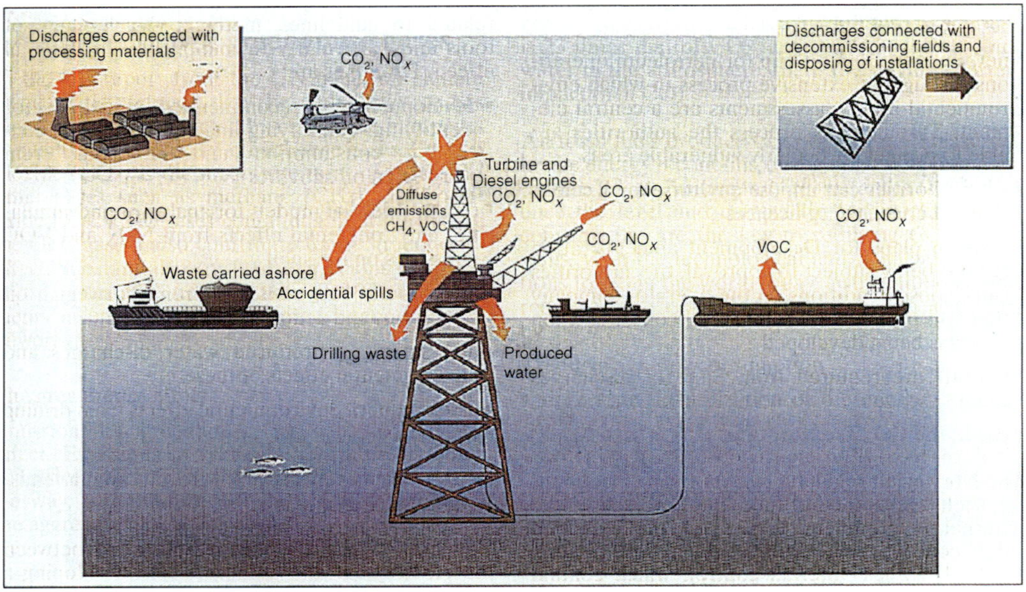

Figure 11.3: The most important environmental impacts from oil and gas operations (from MILJØSOK, 1996).

- chemicals (from drilling operations and production activities);
- drill cuttings (usually embedded in drilling fluid).

Due to regulations and strict discharge limits applicable in Norway, the discharges of oil-based drilling mud have been practically eliminated.

With progressing depletion of oil reservoirs, the produced wellstreams contain increasing amounts of produced water. As a result, produced water has become the main source of oil discharges to sea. Such discharges can be minimized by produced water management techniques, which may include:

- minimizing produced water volumes close to source;
- produced water re-injection for pressure maintenance;
- environmentally friendly chemicals for scale inhibition, corrosion control, and flocculants;
- produced water treatment.

Additionally to the dissolved and dispersed oil, produced water also contains a certain amount of heavy metals and organic compounds. Their relative percentage varies considerably from field to field.

Chemicals are used both in drilling operations and during production. Discharges of chemicals in Norway are strictly regulated by the discharge permit granted by the Climate and Pollution Agency. As a result of the strict regulations, the majority of the drilling additives presently used are inert inorganic materials and degradable organic compounds. Discharges of environmentally hazardous discharges occur in small amounts, as they are found as added compounds and pollutants in products.

Water-soluble production chemicals may follow the produced water discharge, whereas chemicals soluble in oil follow the produced oil stream. Consumption of production chemicals differs much between fields, as determined by reservoir conditions and process facility characteristics.

Cuttings with water-based fluid may be discharged. Oil-based drill cuttings are not discharged, and are either re-injected or transported to onshore treatment facilities.

11.5.2 Impact assessment of discharges to sea

The impacts from discharges are highly dependent on the recipient characteristics. Densely populated and highly industrialized nations enclosing the North Sea have a major influence on the sea through several major rivers flowing into the sea with pollutant from a much wider area, as is similarly the case in other oceans and seas. Oil firms active in the North Sea do not pursue an aim of zero discharge, which would be highly impractical, but they do pursue the aim of zero harm from discharges.

The volume of produced water discharged to sea varies considerably both with the type of production (oil or gas) and throughout the lifetime of a field. The complexity of produced water makes it difficult to assess precisely its environmental impact. Possible impact from discharged chemicals depends among others on their concentration, toxicity, persistence, accumulation potential, and degradation. Impact will also be linked to the recipient characteristics, and has to be evaluated in each specific case.

Dilution calculations for the produced water plume are a valuable tool for evaluating the potential environmental impact. They are often used for EIA. Studies have found that acute toxic effects tend to occur close to the discharge point, and that produced water constituents and production chemicals may cause long-term effects of accumulation and magnification.

Barite contains heavy metals, but apart from this, only small amounts of heavy metals are discharged. Such discharges tend to be in the form of almost insoluble salts, and the biological availability of these heavy metals is therefore very small.

Discharges of added compounds, for example, alkyl phenol ethoxylate, are of great concern to the regulatory authorities. The reason for concern is that alkyl phenols have effects similar to hormones on biological organisms. Alkyl phenols are now practically phased out in the offshore industry in Norway.

Drilling activities generate two major waste categories: cuttings and spent drilling fluid. Discharged cuttings with water-based drilling fluid are shown to give only transient effects on the seabed in a small area close to the discharge point.

11.5.3 Emissions to air

Gases generated by petroleum production include carbon dioxide (CO_2), nitrogen oxides (NO_X), non-methane volatile organic compounds (nmVOC), and methane (CH_4). Emission of sulfur dioxide (SO_2) can occur and depends on the sulfur content of the hydrocarbon and diesel fuel.

The main sources for these emissions are fuel gas and diesel engine exhaust, gas flaring, well testing, oil loading, and diffuse emissions. Gas turbine power generation and gas flaring are the predominant sources of CO_2 and NO_X emissions. Offshore loading of crude oil to tanker ships is a major source of nmVOC emissions. Direct hydrocarbon emissions from vents and fugitives are the main source of CH_4 emissions.

Emitted volumes during exploration activities tend to be comparatively small, as opposed to the more intensive production phase, when increased levels of emissions occur in the immediate vicinity of the operation (Figure 11.4).

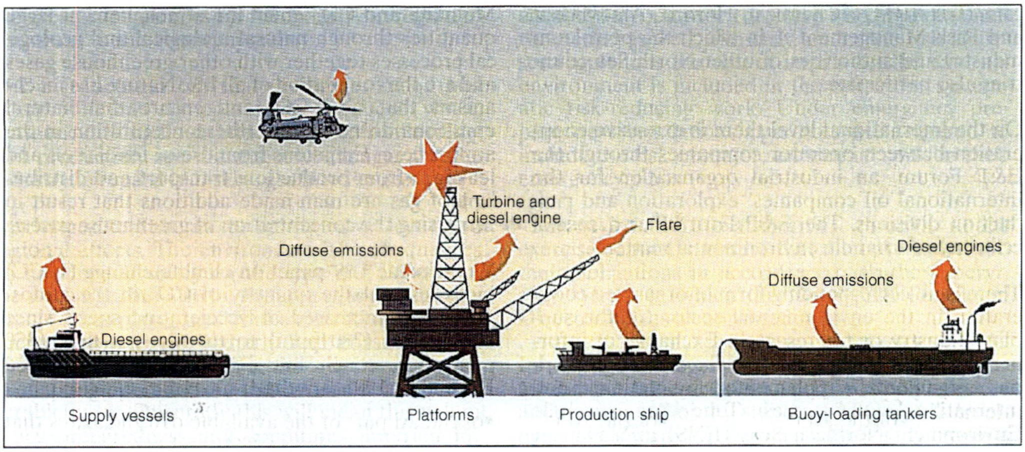

Figure 11.4: Sources of greenhouse gas emissions from Norwegian offshore activities (from MILJØSOK, 1996).

11.5.4 Impact assessment of emissions to air

An important potential effect from emissions to air is acid rain, which may cause acidification of fresh water and soil. It may also cause formation of ground-level ozone, which may be hazardous to plants and humans. Carbon dioxide and methane have global impacts as greenhouse gases. Nitrogen oxide may contribute to regional acidification. Volatile hydrocarbons may cause formation of ground-level ozone.

Sulfur and nitrogen oxides can undergo chemical changes in the atmosphere, forming acidic compounds such as sulfuric and nitric acids, which return to ground as sulfates and nitrates. Acid deposition has a considerable impact on surface waters, soils, and aquifers.

To evaluate the potential for acidification impact related to a field development project, environmental capacity calculations and modeling of precipitation and spreading patterns may be performed for the predicted nitrogen and sulfur emissions. Such calculations indicate whether the planned emissions may increase above the critical load and thereby affect the acidification status of the area.

Ground-level ozone is formed from nitrogen oxides, hydrocarbons (nmVOC), and oxygen in the presence of sunlight. Environmental effects include damage to vegetation, and human health problems may occur as negative influence on breathing capability. Impact assessment of ozone includes the estimation of time of exposure to concentrations above the air quality criteria for ozone.

The severity of impact in an area depends on the amount of emissions together with the area's sensitivity to the emission including recipient characteristics and pre-development conditions.

11.5.5 Principles and methods for reducing pollution

Since produced water is naturally produced with oil, there is limited potential to eliminate the source. There are certain possibilities for reducing the amount of produced water requiring treatment. Re-injection may provide a practical and optimum solution if suitable geological formations are available. The feasibility and environmental impacts of produced water re-injection should be evaluated with regard to reservoir effects (positive or negative), increase in power requirements, and costs. Environmental impacts can be controlled by limiting the use of production chemicals, and by investigating and monitoring the impact of produced water discharge.

The environmental impact of chemicals can be reduced primarily by optimizing drilling operations, process design, and production procedures so as to limit the need for chemicals. Secondarily, the selection of chemicals to be used is made on the basis of their environmental impact.

The use of different drilling fluids was reviewed in Chapter 3. Drill cuttings are a major waste category. Cuttings from drilling with oil-based mud are generally a larger environmental challenge than water-based cuttings. Re-injection is a possibility in some cases. Slim hole drilling and extended use of water-based drilling fluid may also help to reduce the amount of drill cuttings that have to be treated and disposed of.

Efficient use of energy is an important priority in modern petroleum activities for two reasons: the climate gas emissions resulting from energy use and the economic value of the fuels that are spent. Gas flaring has been a common way of disposing of associated gas in many parts of the world, but is increasingly regarded as unacceptable for environmental as well as economic reasons. Gas venting (i.e., release to the atmosphere of natural gas, etc., without combustion) is much more harmful to the global climate than gas flaring, and should therefore be avoided.

Natural gas is often the preferred fuel for generating electricity for the petroleum project's own requirements. It generally has less environmental impacts than other fuels, but such power generation is nevertheless a predominant source of emission of CO_2 and NO_X. Following are some approaches to reduce such emissions:

- energy optimization through heat recovery;
- electric power supply from the grid, if grid electricity is more efficient than can be obtained at field installations;
- combined cycle technology, that is, combinations of gas turbine and steam turbine;
- low NO_X burners;
- reduced emissions of nmVOC.

11.5.6 CO2 injection

Artificial carbon capture and storage (CCS) is a concept for avoiding releases of CO_2 to the atmosphere by capturing CO_2 from large sources such as power plants and petroleum operations, and then storing it permanently. The CO_2 can be stored in geological underground formations, in the ocean at depths of 1000 m or more, or in stable minerals.

CO_2 injection for this purpose is currently (2008) done at a limited number of petroleum fields worldwide, including Sleipner (Figure 11.5) and Snøhvit in Norway and Salah in Algeria. These are gas fields with concentrations of more than 5% CO_2 in the gas wellstreams. The CO_2 is separated from the natural gas at the field installations and re-injected into a separate underground formation.

In several North American oil fields, CO_2 is injected for the purpose of enhancing oil recovery. This CO_2 is usually produced on purpose from underground deposits. The Canadian Weyburn field injects CO_2 obtained as a waste product at an industrial processing plant. The possibility of injecting CO_2 for enhanced oil recovery has been studied also for several North Sea fields, but has not been implemented (as of 2008) due to combinations of reservoir and supply issues.

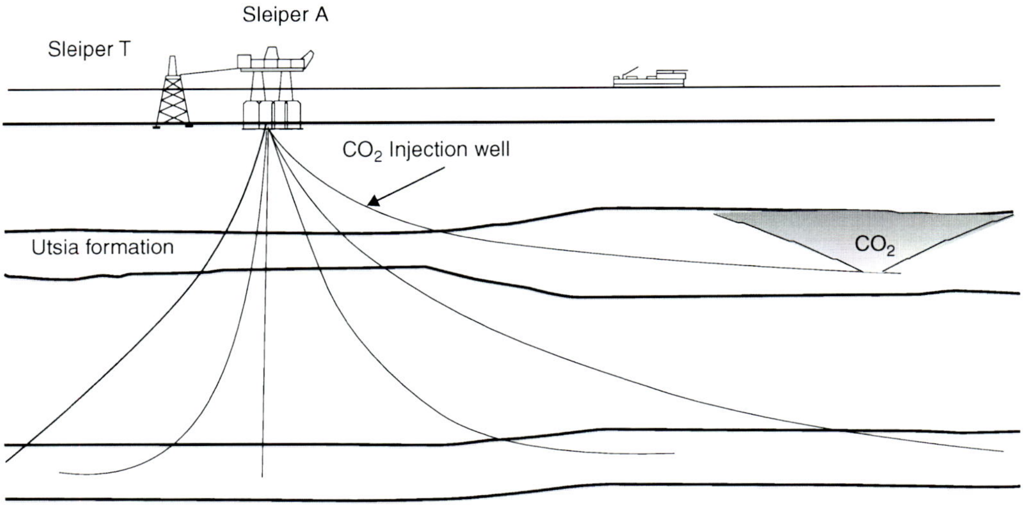

Figure 11.5: Gaining experience with CO_2 injection at Sleipner.

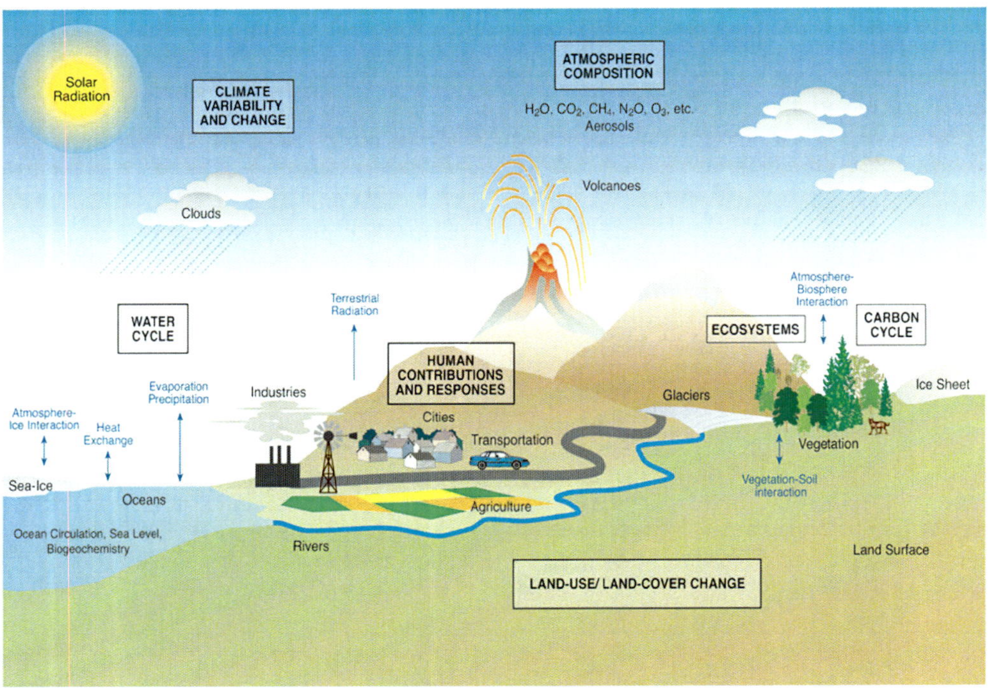

Figure 11.6: Major components needed to understand the climate system and climate change (http://www.climatescience.gov).

11.5.7 Energy and the global climate

Acting on the tentative conclusion that climate change is a serious problem, most nations of the world signed the Kyoto Protocol in 1997. Under the protocol, the developed nations undertook obligations to curb their emissions of climate gases (including CO_2) at certain levels related to the actual levels in 1990. The limitations apply to the 2008–2012 period. There were provisions for emissions trading, that is, countries struggling to meet their limitations can procure permits to exceed them from countries that will emit less than their limitations. Major components needed to understand the climate system and climate change are shown in Figure 11.6.

The European Union has introduced a system for emission permits and trading for climate gases. Firms that emit large quantities of CO_2 and other climate gases, especially power producers, receive permits to do so from the authorities. Surplus permits can be sold, and firms which will emit more than their allocated permits need to buy more of them. Emission permits are traded on several exchanges.

The use of fossil fuels (coal, oil, and gas) is the largest source of man-made emissions of climate gases globally. It is common to link the problem of emissions to the use of such fuels, rather than their production. The climate issue therefore affects petroleum producers indirectly, since it concerns the final applications of the petroleum, but also directly, since significant amounts of oil and gas are also used in the processes of producing them.

11.6 Environmental risk management

Operators on the Norwegian Continental Shelf are required by regulations to establish acceptance criteria for the environmental risk of their activities, and to carry out environmental risk analysis (ERA).

The risk analysis is designed to evaluate the environmental effect, the probability, and consequences of accidental oil spill at sea. The first phase of the analysis is to define the area affected (influence area) through oil spill simulations, and to assess the sensitive biological communities in the area of influence. The second phase of the analysis identifies the dimensioning events and provides probability calculations of the events. The consequences of the dimensioning events are evaluated, and the environmental risks calculated and compared to the acceptance criteria. If the calculated risks are outside the acceptance criteria, cost-effective proposals for risk-reducing measures are established and the analysis is repeated. This will continue until the acceptance criteria are met, resulting in a concept that can be approved. Figure 11.7 illustrates this process as a flow diagram.

The risk assessment work has to be used early in the planning process of operations and developments, making it a genuine decision support tool. In order to define technical solutions representing the best environmental practice, the ERA should be carried out in close cooperation with the operational planners and executers. Not only the calculated results but also the ERA process itself can provide important insights influencing development design.

11.6.1 Acceptance criteria for environmental risk

Environmental damage refers to negative impact of oil pollution affecting biological organisms which may be lethal (e.g., "loss of life") or non-lethal to the exposed biological organism. From this point of view, environmental damage is an incident which disturbs the natural ecological stability outside its natural variability.

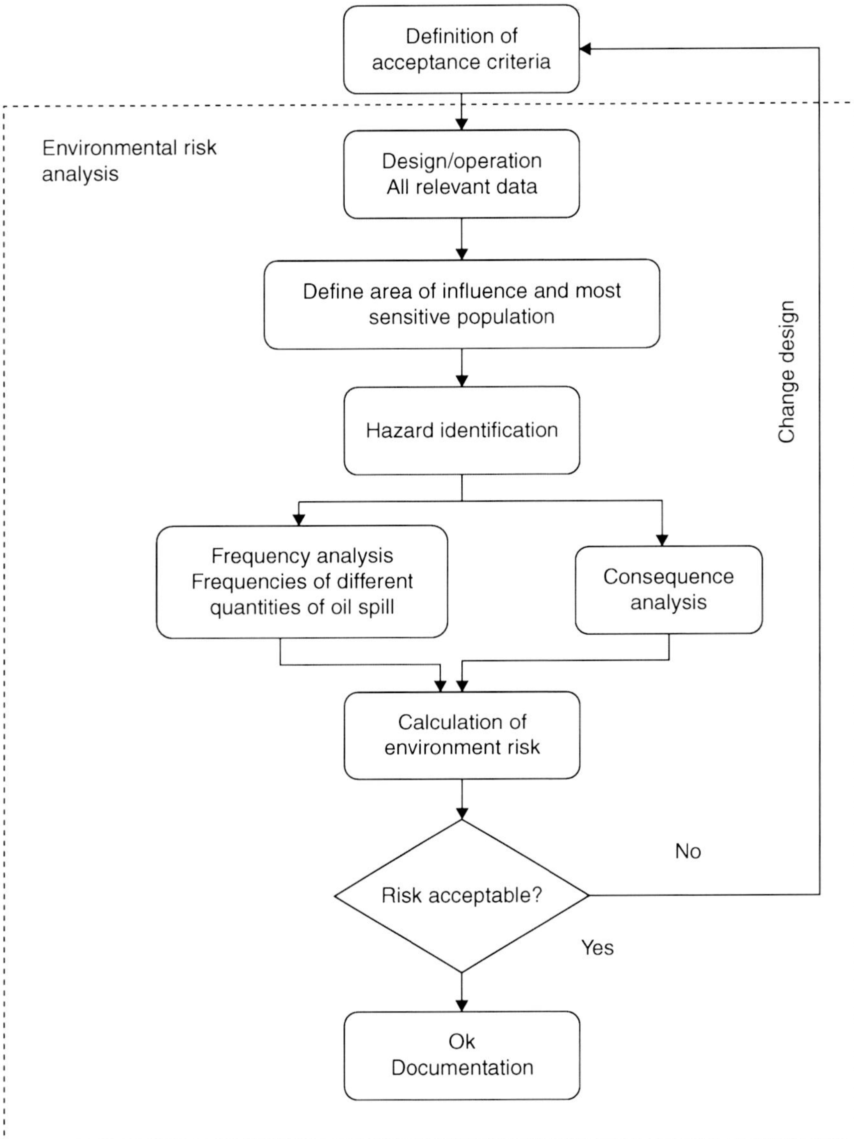

Figure 11.7: Flow diagram for environmental risk analysis.

When assessing environmental damage, it is necessary to have a set of acceptance criteria to evaluate if the environmental risk is acceptable. The main principle adopted on the Norwegian Continental Shelf for establishing such criteria is that the environment shall be in a "natural state" most of the time, meaning undisturbed: the duration of any environmental damage shall be insignificant in relation to the expected time between such damages.

According to NORSOK standard Z-013 (the NORSOK standards is a set of industry-agreed standards applicable to Norwegian petroleum installations), environmental damage is defined as the direct or indirect reduction of one or several resources resulting from an accidental spill. The recovery time for at least one of the affected resources must be at least 1 month for the effect to be classified as environmental damage. As a single operator cannot control the total exposure, assumptions are made regarding an average activity level (number of oil fields, installations per field, and operations per year) in different areas on the Norwegian Continental Shelf.

To calculate the acceptance criteria, the operator has to decide which value to be interpreted as "insignificant." If the "insignificant" is defined as 5%, this means that the most sensitive population can as a maximum be disturbed in 5% of the time. This implies that a damage which is recovered within half a year in average is acceptable if the calculated risk frequency for the damage is less than once per 10 years. (Note, however, that this is an acceptable *risk*; the operating companies do not accept any environmental damage.)

11.7 Oil spill contingency

Oil spills are a major environmental hazard associated with oil production. Highest priority should always be placed on the prevention of spills. In addition, proper emergency management to handle any spill is required. Oil spill response planning is one important part of this contingency planning activity.

11.7.1 Contingency planning

Reference is made to the NORSOK standard Z-013 "Risk and emergency preparedness analysis."

The successive steps in the contingency planning may be summarized as follows:

- data collection and risk assessment;
- deciding on response objectives;
- choosing response options;
- equipment and supplies.

Sensitivity mapping and analysis of oil spill drift and spreading are essential inputs for the oil spill contingency planning.

11.7.2 Response strategies

Area, site, and season-specific conditions may represent constraints to the response options, including oil recovery and clean-up techniques. Acceptable response options at particular locations must be established with the relevant authorities.

In general, the oil spill combat strategies may be subdivided into three main categories:

- mechanical recovery;
- chemical recovery;
- "let the nature do the job."

Mechanical recovery contains the strategy to use booms and skimmers for containing spilt oil offshore, and the use of shovels and diggers to contain the oil onshore.

Chemical recovery is the use of chemicals to help disperse the oil at sea, and to assist cleaning the shoreline. The effectiveness of chemicals is due to their enhancement of the natural forces and acceleration of the physical break-up of an oil mass, causing it to be dispersed as small droplets into the water column.

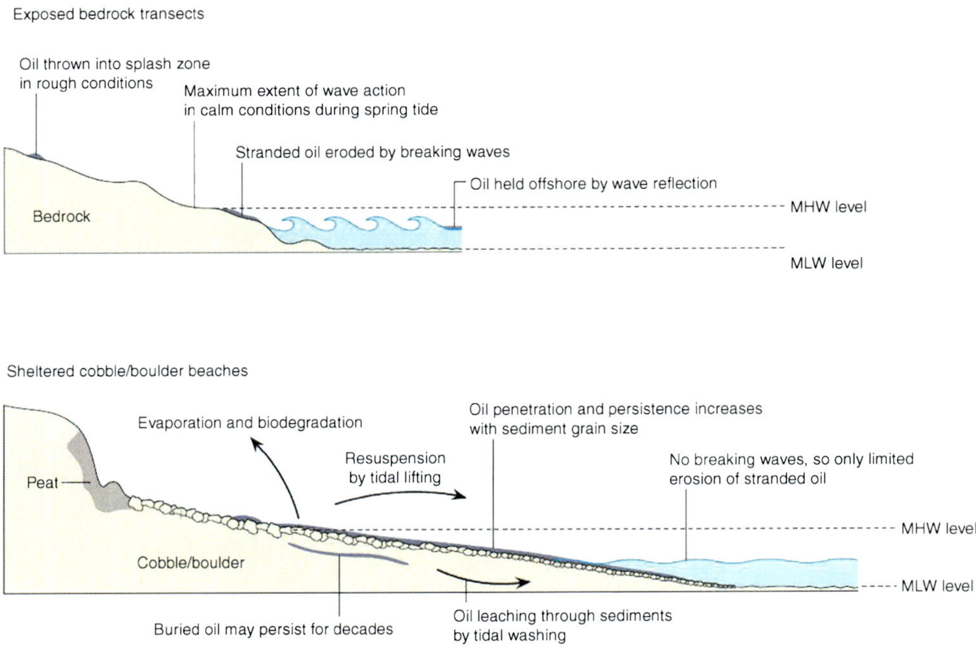

Figure 11.8: Fate of stranded oil on exposed and sheltered rocky shores (from the homepage of IPIECA, http://www.ipieca.org/index.php, IPIECA, 1996).

"Let the nature do the job" implies that there are scenarios when a choice of no active response except for monitoring and reporting might be the most appropriate response.

In a high proportion of incidents, a combination of different methods is applied. The chosen strategy should reflect the primary response objective. Decisions on the response objectives must take the inherent equipment limitations and realities of the situation into account.

In Norway, the use of dispersants has not been favored, the priority method and strategy being use of mechanically spill recovery units as close to the spill site or source as possible. Dispersant may be a feasible supplement in particular cases, but this option needs to be evaluated on a case-to-case basis.

Whatever the case, it is essential that the response options are carefully considered and agreed in advance by all relevant parties. For all sea and shore areas, they should be specified in the contingency plan (Figure 11.8).

11.7.3 Organization of the Norwegian oil spill emergency preparedness

Norwegian oil spill emergency preparedness consists of three contingency levels:

- national government;
- municipality;
- company.

Their different responsibilities are outlined in regulations. The Climate and Pollution Agency is responsible for the administration and operation of the national oil spill preparedness, and ensures that adequate municipal and private emergency preparedness are in place. The state's responsibility is mainly to limit pollution from ships and from sources that are not covered by private contingency measures. Companies (operators) are required to have prepared contingency measures sufficient to handle any spill that can emerge from their operations.

In the offshore petroleum sector, the operating companies' emergency preparedness is administrated by an industry association set up for this purpose (NOFO – Norwegian operators' association for oil protection). This organization supports operators in complying with the applicable requirements for oil spill contingency. In the event of a spill, NOFO will mobilize the necessary equipment, and the operator will organize the spill response based on the contingency plan in place.

NOFO has developed emergency contingency plans for the mobilization and operation of jointly owned resources. The organization has established bases comprising equipment, personnel, and vessels, which are kept in a state of readiness. In the case of an oil spill, each member company of NOFO is entitled to utilize the organization's resources as part of its own overall preparedness. This enables the operating companies to respond to a spill at short notice.

Should an oil spill reach the coast, local and national authorities would combat the spill in cooperation with the oil company involved.

11.8 Summary

Authorities and the general public require petroleum operations to be environmentally sustainable. Some impact on the environment may be inevitable, but need to be kept to a minimum based on systematic management and monitoring. This chapter has reviewed requirements and practices that are applied in Norwegian offshore petroleum operations but many of which can have wider applications.

Before petroleum operations in an area commence, the existing environmental conditions should be carefully documented with an environmental baseline survey. International standards and guidelines have been developed for managing environmental protection. Regulations and methods for environmental protection are related to those for safety management, since the need to avoid accidents is a common issue. Efficient use of energy in petroleum operations is both an environmental and economic issue, due to the global challenge of climate change as well as high energy prices. Protection against oil spills is a particularly critical task in the environmental management of petroleum operations.

References

Environment & Resource Technology Ltd. 1997. *The Faroe Islands. An Environmental Statement.* Jointly published by the Atlantic Margin Group (AMG), Statoil Efterforskning og Produktion A/S, Enterprise Oil plc, Mobil North Sea Ltd., ISBN: 87 986247 0 9.

Gudmestad, O.T., Zolotukhin, A.B., Ermakov, A.I., Jakobsen, R.A., Michtchenko, I.T., Vovk, V.A., Løset, S. & Shkhinek, K.N. 1999. *Basis of Offshore Petroleum Engineering and Development of Marine Facilities with Emphasis on the Arctic Offshore.* Oil and Gas Printing House, Gubkin University, Moscow.

ISO, International Organization for Standardization 1996. *International Standard. Environmental Management Systems—Specification with Guidance for Use.* ISO 14001:1996, Geneva, Switzerland.

IPIECA, The International Petroleum Industry Environmental Conservation Association 1996. *Biological Impacts of Oil Pollution: Rocky Shores.* Volume 7, available at: www.ipieca.org, London, UK.

MILJØSOK. 1996. *The Oil Industry Aassumes Responsibility.* Report from the MILJØSOK Steering Group, December, available at: www.olf.no.

Ministry of Petroleum and Energy, Oslo, Norway. 2005. *Environment 2005, The Norwegian Petroleum Sector.* ISSN 1502-3400 (English editor Vanja Stake).

Ministry of Petroleum and Energy, Oslo, Norway. 2006. *Facts, The Norwegian Petroleum Sector 2006.* ISSN 1502-5446, pp. 59–77.

UNEP. 1994. IE/PAC *Technical Report 37.* E&P Forum (OGP) Report 2.72/254. ISBN 92-807-1639-5.

Wood, E., Hoagland-Grey, H. & Smith, J. 1996. *The West Coast of Ireland. An Environmental Appraisal.* Dames and Moore/OPRU, London/Pembroke. Published jointly by Texaco, Enterprise Oil PLC, Kerr-McGee Oil (UK) PLC, and Statoil (UK) Limited.

Licensing and fiscal regimes

This chapter concerns the terms under which a petroleum firm may obtain rights to explore for and produce petroleum.

The *licensing system* for petroleum rights in a country is the set of laws, regulations, and agreements by which rights for petroleum exploration and production may be granted to commercial entities.

The *fiscal regime* for petroleum is the set of laws, regulations, and agreements which govern the transfer of economic benefits between commercial entities and the nation.

The fiscal regime of a country can be considered part of the licensing system, or at least closely linked with it.

In this chapter, basic types of licensing and fiscal regimes will be described and discussed, as well as their implications for the economics of the parties. The first three sections introduce the legal basis, licensing regimes, and the issue of joint interest. The remaining sections of the chapter introduce fiscal regimes and discuss certain key issues related to them.

12.1 The legal basis: ownership to resources in the underground

The licensing regime concerns the question: how can a commercial entity acquire rights to find and produce petroleum in a country? First, this requires clarification of another question: who holds the rights to petroleum in the underground in the first place?

National legislation is needed to determine the entitlements to resources in the underground. At least two systems are possible:

- Whoever owns the ground above, also owns the resources below.
- The state owns the resources below, regardless of ownership to the ground above.

The former system is found in the United States (with variations between the states). The latter system of state ownership is found in many other countries. It is also common for the state to have ownership to resources below offshore waters.

In the case of Norway, the state has property rights to underground petroleum deposits both offshore and onshore as established in two acts:

- Concerning offshore petroleum deposits, in the Petroleum Act from 1996, clause 1-1 (translation from the original Norwegian text): "The Norwegian State has the proprietary right to subsea petroleum deposits and the exclusive right to resource management."
- Concerning onshore petroleum deposits, a similar clause in an Act from 1973. As it happens, no petroleum has been found onshore in Norway, but state ownership to any such petroleum is nevertheless established by law.

Following from the above acts of law, it is the privilege of the state to grant any rights to petroleum extraction to any party, and no party can undertake such activities without the consent of the state.

For the purpose of this text, it is presumed that the state initially holds such proprietary rights to petroleum in the underground, as is indeed the case in most countries of the world (including Norway, as shown above).

In countries that have a federal structure, such as the United States and Nigeria, the rights of the "state," as referred above, may reside either with the national (federal) government or with the governments of the constituent entities, which may be called provinces, states, republics, etc. This issue of national power distribution also needs to be clarified by legislation.

12.2 Licensing regimes: concessionary and contractual systems

It is common to distinguish between two main types of licensing: concessionary systems and contractual systems.

In a *concessionary system*, the state grants to a commercial entity an exclusive right to extract petroleum. Such rights within a defined area are referred to as a *license*, and the commercial entity (oil company) that holds the license, is a *licensee*. The licensees will own the installations they put in place as well as the petroleum that they obtain. The licensees have to provide certain financial benefits to the state in the form of royalties, rentals, taxes, etc. These benefits will be discussed further below as part of the fiscal regime. A concessionary system may also be known as an *equity system* or *royalty/tax system*.

A *contractual system* is based on the premise that the petroleum belongs to the state and is produced for its benefit. The exploration and production activities are carried out by a commercial entity (oil company), which provides the necessary equipment and personnel, and receives compensation usually in the form of a share of the petroleum obtained. The oil company is then referred to as a *contractor*, operating under a production sharing contract (or other form of agreement) with the state. The party that represents the state's interest in such a contract is in many cases a state-owned oil company if such exists in the country, otherwise a ministry of the government.

There are several variations of contractual systems. The most common form is *production sharing contracts*. Here the contractor is compensated in kind, that is, by receiving a share of the oil and/or gas produced. The share of production received by the contractor has two components: recovery of costs ("cost oil") and a share of the remaining surplus ("profit oil"). In addition, the contractor may be required to lift and sell the state's share of petroleum on its behalf.

An example of a list of articles for a production sharing contract is shown in Table 12.1.

In *service contracts*, the contractor is compensated by payment in cash. A distinction is made between *pure service* contracts, in which the contractor receives compensation independently of the profitability or size of the developed resources, and *risk service* contract, in which the contractor's compensation is somehow linked to the value of the developed resource. Some contracts also include rights for the contractor to purchase the oil at defined terms (*buyback contract*). A production sharing contract or a service contract can also be a *technical assistance contract*, in which a company is engaged in order to carry out a project at an existing field, such as rehabilitation, redevelopment, or enhanced oil recovery.

Service contracts are often regarded as suited to situations where a known field needs development work in order to start or improve oil production, that is, there is little or no exploration risk involved. Production sharing contracts are then better suited to areas in which production depends on exploration success to be achieved, because such contracts tend to allow a substantial profit for the company in case of success and as a reward for the exploration risk taken. Table 12.2 lists the predominant licensing systems in selected countries.

Table 12.1: List of articles of a production sharing contract (example).

I	Definitions
II	Annexes to the agreement
III	Grant of rights and term
IV	Work program and expenditures during exploration period
V	Mandatory and voluntary relinquishments
VI	Operations after commercial discovery
VII	Recovery of costs and expenses and production sharing
VIII	Title to assets
IX	Bonuses
X	Office and service of notices
XI	Saving of petroleum and prevention of loss
XII	Customs exemptions
XIII	Books of account: accounting and payments
XIV	Records, reports, and inspection
XV	Responsibility for damages
XVI	Privileges of government representatives
XVII	Employment rights and training of national personnel
XVIII	Laws and regulations
XIX	Stabilization
XX	Right of requisition
XXI	Assignment
XXII	Breach of agreement and power to cancel
XXIII	*Force majeure*
XXIV	Disputes and arbitration
XXV	Status of the parties
XXVI	Local contractors and locally manufactured material
XXVII	Language
XXVIII	General
XXIX	Approval of the government
Annex "A"	Boundary description of the concession area
Annex "B"	Illustrative map showing area covered
Annex "C"	Charter of operating company
Annex "D"	Accounting procedure
Annex "E"	Map of the national gas pipeline grid system

Table 12.2: Licensing systems around the world.

	Concessions	Production sharing contracts	Service contracts
Far East	Australia, Brunei, South Korea, New Zealand, Pakistan (On), PNG, Thailand, Timor	Bangladesh, Cambodia, China, India, Indonesia, Laos, Malaysia, Mongolia, Myanmar, Pakistan (Off), Timor, Vietnam, Nepal, Sri Lanka	Philippines
Former Soviet Union	Kazakhstan, Russia	Azerbaijan, Georgia, Kazakhstan, Kyrgyzstan, Russia, Turkmenistan, Uzbekistan	
Latin America	Argentina, Bolivia, Colombia, Costa Rica, Paraguay, T&T (On)	Belize, Cuba, Guatemala, Guyana, Jamaica, Nicaragua, Panama, T&T (Off), Uruguay	Brazil, Chile, Ecuador, Haiti, Venezuela, Honduras, Panama, Peru
Middle East	Abu Dhabi, Ajman, Dubai, Fujairah	Bahrain, Jordan, Libya, Oman, Qatar, Syria, Yemen	Iran, Kuwait (OSA), Saudi Arabia; Iraq
North America	Canada, United States		

12.3 Partnerships: several companies in one project

In both concessionary and contractual systems, it is possible for several companies to participate as partners in one license or contract. Oil companies often prefer such partnerships as a way of spreading their investments and risks across several projects. Governments may also prefer it in order to get alternative viewpoints on the development, and as a safeguard against excessive costs. There are several ways in which such partnerships can be formed:

Partnerships initiated by the government: In some countries having a concessionary system (such as Norway), the government may decide to award a license to several companies jointly, even after they applied separately for the license.

Partnerships initiated by the companies: Several oil companies may elect to approach the government jointly to obtain a license or contract, and will form a partnership for that purpose.

Farm-in: A company which initially held an entire license or contract, or strike *large*, may agree that another party will enter the project as a new partner.

In a jointly held license or contract, an agreement between the parties is required to determine how benefits and responsibilities will be shared between them. This is called a *joint operating agreement*. For partnerships initiated by the government, the government also provides a standard joint operating agreement which it requires the partners to sign as a condition to receiving a share in the license.

One of the participating firms is designated as the *operator*, while the other firms are *non-operating partners*. The operator carries out the actual operations using his personnel and contracted services. The partnership as such is usually not incorporated or recognized as a legal entity. The non-operating partners reimburse the operator for their shares of all expenditures. Each partner lifts its respective share of oil and gas produced (to the extent not accruing to the government). Investments and other major decisions are made according to rules set in the joint operating agreement.

12.4 The fiscal regime and its components

The fiscal regime consists of the various taxes, royalties, fees, production shares, and any other monetary benefits which shall accrue to the state from the petroleum activities. This is determined by the applicable legislation of the country as well as negotiated agreements between government and commercial entities. From the perspective of the participating firms, the fiscal regime defines the value that they will give up for the benefit of the state. Johnston (1994) provides a more comprehensive overview of international fiscal regimes

The fiscal regime of a country for petroleum activities usually contains several, but not all, of the following components:

- *Signature bonus*: A sum of money that the licensee or contractor is required to pay on being granted the license or contract. It is common in contractual license systems, and also (sometimes under different names) in concessionary systems. Some countries request large signature bonuses (many million US dollars) which will be individually negotiated for each license. Some other countries require no signature bonus, but may require a modest fee intended only to cover the administrative costs of handling applications. A signature bonus provides the government with a financial benefit from granting a license or contract, irrespective of whether there will be any production of petroleum.
- *Area rental fee*: An amount payable per year and per square kilometer (or square mile) of surface area covered by the concession or contract. An area rental fee tends to motivate oil firms to relinquish any part of the licensed area where they have concluded that prospects for finding petroleum are poor, subject to provisions set forth in the agreement.
- *Royalty*: A percentage share of produced oil and/or gas which the government shall be entitled to receive. The agreements have provisions for whether royalty will be taken in kind or in cash. If taken in kind, the government receives the actual product, and must arrange for it to be handled in terms of logistics and disposition. If taken in cash, the government receives the monetary value equivalent of the royalty percentages of production, and the agreement must then contain provisions for how to determine this value. Royalty can apply under both concessionary and contractual systems. It can apply to oil as well as gas.
- *Production share*: This applies to contractual systems, and more specifically to production sharing contracts. The contractor and the government will share the oil which is produced, after allowing some oil to be set aside for the contractor to cover his costs. This part of production is referred to as cost oil, cost recovery petroleum, or some similar term. The part of oil production that remains after deducting cost oil as well as any royalty is referred to as profit oil or some similar term. Production sharing contracts contain a table specifying how this oil will be shared. It is common in such contracts to have a scale determining that the government's share of profit oil will be larger if the resources turn out to be large and valuable by some measure.
- *Corporate income tax*: In most countries, oil companies have to pay income tax to the country in which they produce oil and gas, on similar basis as other business firms operating

there (a few countries either apply no corporate income tax at all or exempt income from production sharing contracts from income tax). Corporate income tax is generally paid on the firm's pre-tax profit, that is, income from all its activities in a country after deduction of all expenses, depreciation, finance costs, etc., in accordance with accounting rules set forth in the country's general tax legislation.

- *Special petroleum tax*: This type of tax is mainly relevant under concessionary systems, where it serves a similar purpose as production sharing under a contractual system. As an example, in Norway, oil firms have to pay a special petroleum tax of 50% on top of the ordinary corporate income tax of 28% (the special petroleum tax in Norway is levied on the pre-tax profit, as is also the corporate income tax, but some additional deductions apply to the special petroleum tax). The special petroleum tax is intended to capture a large part of the "resource rent" for the nation's benefit, that is, the extra profit that derives from the extraction of limited natural resources.

Figure 12.1 outlines the allocation of gross revenues from oil production under a production sharing contract. The revenues collected for the state are jointly referred to as the government take. It includes the royalty on gross production, the state production share of profit oil, and the tax on corporate income. Not shown in the figure are a signature bonus and an area fee, which apply in many countries. Included in the figure is a cost element, "costs outside contracts," which would include costs that the firm incurs in developing its business in the country, but which cannot be assigned to any particular contract. Such costs may be deductible for income tax purposes, depending on the applicable legislation.

Examples of how profit oil can be shared in a production sharing contract are given in Tables 12.3 and 12.4. In the first example, different shares for the government and the contractor are triggered by rates of production, which is calculated as a quarterly average. The shares apply cumulatively, so that the contractor keeps 40% of the first 27,000 barrels per day, but smaller shares of any production above 27,000 barrels per day. It is the gross daily production that triggers the different percentages, but the amounts of oil to be shared will be whatever is left after cost recovery.

The example in Table 12.4 shows production sharing based on the *R*-factor. This factor is calculated at any time as the ratio of the contractor's cumulative receipts to his cumulative

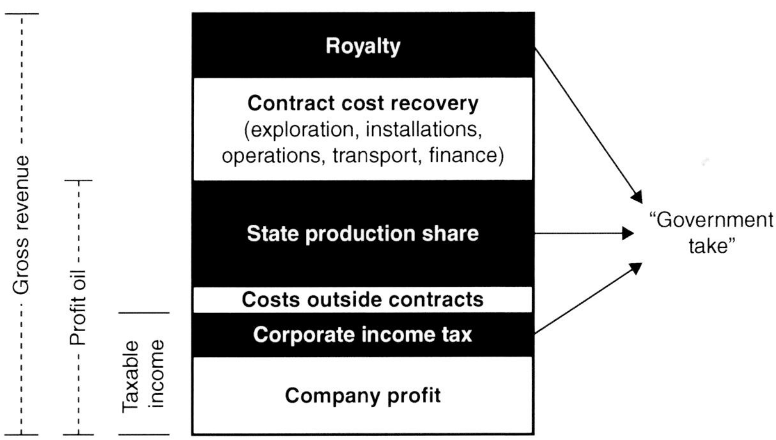

Figure 12.1: Allocation of revenues under a production sharing contract.

Table 12.3: Example of production sharing table based on daily production of oil and gas.

Production from contract area	State company share (%)	Contractor share (%)
Crude oil—daily production		
Up to 27,000 barrels per day	60	40
27,000–53,000 barrels per day	65	35
53,000–107,000 barrels per day	70	30
107,000–160,000 barrels per day	75	25
Above 160,000 barrels per day	80	20
Natural gas and LPG—daily production		
Up to 150 MMSCFD[a]	60	40
150–300 MMSCFD	65	35
300–600 MMSCFD	70	30
600–900 MMSCFD	75	25
Above 900 MMSCFD	80	20

[a] MMSCFD, million standard cubic feet daily.

Table 12.4: Example of production sharing table based on *R*-factor.

R = (cumulative revenues)/(cumulative expenditures)	State company share (%)	Contractor share (%)
$R < 1$	65	35
$1 < R < 1.5$	70	30
$1.5 < R < 1.75$	75	25
$1.75 < R < 2$	80	20
$R > 2$	85	15

expenditures since its start in the contract area. When *R* reaches 1, the contractor's cumulative receipts equal his cumulative expenditures, that is, he has recovered his cost without any profit. The example in Table 12.4 shows the contractor receiving 35% of profit oil up to that point, and declining shares later.

The petroleum activities incur costs which must be recovered for the operations to be economically viable. Costs directly attributable to the production operations can be recovered under provisions of the production sharing contract. The portion of revenues that is left for the company after deducting all project-related costs and the various fiscal elements is sometimes referred to as "company take."

12.5 Cost recovery

Where several parties will share in a petroleum project, it is necessary to determine the relevant costs, as they impact on payments that are to be made between the parties. Costs are dealt with differently in the various types of arrangements for petroleum developments.

In situations with production sharing and service contracts, three different cost recovery procedures may operate simultaneously:

- Where several oil firms are partners in a contract or license, the operator will recover expenditures from the other partners according to their shares in the project.
- Production sharing contracts and service contracts have provisions for the oil firms to recover their costs.
- For the purpose of corporate income tax, costs can be deducted from taxable income based on tax accounting rules.

Each of these is reviewed below.

Expenditure and *cost* are two different concepts: expenditure is a payment made. A cost is a consumption of value. The two are often, but not always, the same for accounting purposes. For instance, wages paid to operating personnel during a month are expenditure in that month as well as a cost in the same month. However, if a machine is bought for $1 million in the year 2010 which will operate for 10 years (2011–2020), then this is an expenditure of $1 million in 2010, but not a cost during that year. The cost will instead be $100,000 for each of the following 10 years, appearing as depreciation in the financial statements (this is a simplified example; applicable accounting procedures may result in a different depreciation schedule). A capital investment is expenditure but not a cost. Depreciation is a cost but not expenditure. The difference between expenditure and cost is a matter of timing: the $1 million expenditure for the machine in 2010 will be accounted for as costs eventually, but only after 10 years.

12.5.1 Cost recovery between operator and non-operating partners

Where several oil companies participate as partners in a petroleum project, they need to have agreed on how costs will be shared. The common arrangement is immediate reimbursement of all expenditures within the project according to each partner's share.

As an example, if companies A and B are each 50% partners in a license, with A being the operator, then A will procure and pay for all goods and services needed in the exploration, development, operations, and decommissioning of the project. A will issue periodic statements to B of all such expenditures (sometimes referred to as joint interest billing), and B will forward cash to A corresponding to B's 50% share of the project. B will forward cash simultaneously with the required expenditures, so that in principle A should neither lose nor gain any interest or other value from being an operator rather than a non-operating party.

The operator A also includes the salaries, etc., of his own personnel in the expenditures to be shared by B. But A's personnel may be working on several projects. A therefore needs to have procedures for allocating personnel costs, as well as general overhead and administrative costs, between the different projects that the company is engaged in. Personnel with operating companies therefore usually write time sheets to record to what extent they have worked on projects that other parties will also pay for. The operator A can also recover expenditures for support work, etc., taking place outside the country of the actual operations, based on reasonable documentation that such expenditures were made in the interest of the joint project.

12.5.2 Cost recovery under production sharing contracts and service contracts

Under production sharing contracts, the issue of cost recovery influences how oil is shared between the government and the contracting oil company. As indicated in Figure 12.1, a portion of the produced oil is used to recover costs, before the remaining "profit oil" is shared between the government and the contractor. The types of costs that can be included in the cost recovery are specified in document (often called an *accounting procedure*) that is annexed to the production sharing contract.

Under production sharing contracts, cost recovery is generally based on expenditures, that is, capital expenditures can be recovered immediately (to the extent there is available revenue) without calculating depreciation.

Costs are often larger in the early stages of oil production than they will be later, not least because some costs have accumulated prior to the start of oil production. In order to ensure that the government gets at least some value of oil production from the start, many production sharing contracts have a maximum limit for cost recovery, expressed as a percentage of oil production. If, as an example, this limit is set at 60%, then no more than 60% of the oil produced in a year can be used for cost recovery (as "cost oil"), and at least 40% will then be subject to production sharing between the government and the contractor. Any such limit would be specified in the production sharing contract, and may of course be something else than 60%; some production sharing contracts have no such limit, i.e., the limit is then 100%. If the contractor has incurred expenditures which should be recovered but are in excess of the 60% limit, he is normally allowed to carry these excess costs forward for recovery in a later year, and may be entitled to add interest to such costs carried forward.

Interests and other finance costs are normally excluded from cost recovery. The implication of this is that firms must cover the costs of finance out of their profit share.

Service contracts often have similar provisions for cost recovery, but contract provisions must be reviewed individually.

12.5.3 Cost deductions for income tax purposes

Most countries have tax legislation under which companies (as well as individuals) have to submit a statement of their income at the end of the year, which becomes the basis for taxation. The tax legislation normally applies to companies in general—whether or not they are oil companies. Typically there is a flat tax rate on the company's pre-tax profit (the company tax rate is 28% in Norway, and close to that rate in many other countries). The pre-tax profit is essentially income minus all costs, but the tax legislation includes provisions for how costs may be calculated for that purpose.

Income calculation for company taxation is normally based on costs, rather than expenditures, being deductible from revenues. This is a difference from the calculation of cost recovery for production sharing contracts, which are based on expenditures rather than costs. The main difference is in the treatment of capital expenditures. Whereas they may be deductible directly for production sharing, they have to be calculated as depreciation for tax purposes.

In addition, there may be deductible items in the tax calculation which are additional to any contract cost recovery. This may include costs for general administration, business development, etc., that may not be recoverable under any project contract, but may nevertheless be deductible as business expenses for tax purposes.

It is not unusual for oil companies to conduct business in a country in the name of an affiliate that they have registered elsewhere. Nevertheless, the tax legislation of the country usually includes a provision that such business activities in the country are taxable there even if conducted in the name of a firm that is registered abroad.

12.6 Getting proper value for oil

Petroleum development projects are ultimately driven by the value that can be obtained for the oil and gas. Getting the value right is no trivial issue: international market prices for oil can be readily observed, and are often used to guide oil prices around the world, but they fluctuated between $10 and 145 per barrel during 1998–2008. Getting good value for a particular source of crude oil also depends on understanding and positioning the oil correctly in international markets, including appropriate logistical facilities. For natural gas, this entails added challenges as will be described in the next section.

The participating oil firms are usually in a position to sell their crude directly on world markets, for which they have the relevant capabilities, and are therefore in a position to ensure that they obtain proper value for their oil. A government may be in a different position: the income it will derive from petroleum production depends greatly on the prices that will be assigned to it, but governments generally do not possess the capabilities required to market petroleum professionally.

In a purely concessionary system, the government generally does not get involved in the business of marketing oil and gas. Instead, it relies on accounting procedures to determine the profit from which it will derive tax. For that purpose, government must ensure that the price assigned to such oil reflects its real value—otherwise, an oil company might sell the oil for an artificially low price to its refining affiliate abroad or indeed to its trading affiliate located in a tax haven, in order to avoid tax to the country of production.

Governments have taken different approaches to deal with this issue. The Norwegian government operates a *norm price* system, under which a board of experts determines the fair market value of Norwegian crude oil qualities on a quarterly basis. The norm price applies in the tax calculation, regardless of what prices the oil companies actually invoiced.

Production sharing contracts usually contain provisions for determining how crude oil which is exported by the contractor shall be valued when calculating the fiscal elements. By a common type of provision, the actual realized price applies if the crude oil was sold to a third party (i.e., unaffiliated with the seller), but if the buyer was affiliated with the seller, then there are more elaborate procedures to assign a price based on other observed prices in the region over a certain period. Governments are not always satisfied that this type of provision will ensure that they get the proper value for oil.

Under production sharing contracts, government has another option, which is to handle and dispose of oil and gas itself. This is indeed the logical starting point under such contracts, which are based on the premise that petroleum belongs to the state. Receiving and handling petroleum requires capabilities that government ministries are usually not well equipped to provide. What then can the government do with the oil and gas to which it is entitled?

The problem has at least two possible solutions: the government may leave the oil and gas for the contractor to handle and receive money instead. Or it may form its own oil company to handle the petroleum. Both solutions are found around the world.

If the country has a state oil company, the production sharing contract may be entered into between the contractor and the state oil company, rather than directly with a government ministry as would otherwise normally be the case. In either case, the production sharing contract may include provisions that give the government (or state company) the option of either handling the petroleum itself or to have the contractor market it on its behalf.

12.7 Dealing with gas in fiscal regimes

Traditionally, production sharing contracts have focused on crude oil as the valuable commodity. Gas was either assumed not to be of any value or left as an unresolved issue that would have to be dealt with when and if necessary. Much associated gas was (and still is) flared off as an unwanted by-product of oil production. This practice is increasingly regarded as unacceptable; both for environmental reasons and because natural gas has emerged as a valuable global commodity which can be captured and monetized on economic grounds.

Some production sharing contracts provide for sharing the crude oil, but not gas. Such contracts may make a distinction between associated gas (which arises as a by-product of oil production) and non-associated gas (which can be produced independently of oil). Any associated gas may be used as a fuel in oil field operations, and any surplus was to be delivered to the state oil company if it so required. The associated gas may otherwise be flared subject to permission from the government. Any non-associated gas may be subject to a separate development plan agreed between the parties.

Table 12.3 is an example from a production sharing contract which more explicitly recognizes natural gas as a valuable product, and therefore has a production sharing table for gas as well as oil. At the time this contract was made, the country in question had developed significant domestic use of natural gas, and was preparing to become an exporter of natural gas as LNG. The production sharing contract had provisions that, in the case of gas discovery, the state oil company would try to find domestic outlets for the gas, in which case the intention was for the gas from the project to be sold to the state oil company at a price formula defined in the contract. Alternatively, an export option as LNG could be developed.

Although there can clearly be possibilities for defining the value and disposition of gas in a production sharing contract, this is a more difficult issue than is the case for oil, due to the costly logistical requirements for bringing gas to a market. The possibilities for developing a good solution for gas utilization depend in part on local circumstances, which again have implications for the fiscal provisions.

12.8 Fiscal terms as a negotiation issue

A production sharing contract is usually the outcome of a negotiation between a government (or its agent) and one or more commercial firms. Several of the fiscal elements mentioned earlier may be considered as negotiable in the process of working out the terms for a particular license or contract.

The government may indicate that certain elements can be negotiable, whereas others are considered non-negotiable. Least likely to be negotiable are the tax elements (corporate income tax and special petroleum tax if any), since they are normally determined by tax legislation that is passed by Parliament.

Besides the fiscal components, the negotiations for a production sharing contract or concessionary license are likely to feature several issues which can have implications also for the fiscal terms. Some such issues are indicated as follows:

- *Work obligations*: It is common for the contractors or licensees to undertake certain obligations in terms of exploration work to be carried out, such as an amount of seismic work, exploration wells, etc., and to report the results to the government. This has the function of increasing the knowledge of the nation's petroleum potential, which is in the government's interest.

- *Participation of state oil company*: The government may request an option for the state oil company to join the project as a partner on privileged terms. One possible provision is that the (non-state) partners will carry out the exploration activities, and if commercial oil and gas is found, the state company may opt in, having to pay its share of expenditures from then on but not having to pay its share of past exploration costs.
- *Operatorship and technical assistance*: The government may request that the state oil company will become the operator of any development at some stage, for instance, after the initial development is completed and production has started, and possibly with another partner acting as technical assistant. The purpose of this would be to support the development of the national firm in terms of capabilities on its way to becoming a fully qualified oil company. This aim has to be considered against the danger of having a not fully qualified firm as operator of a complex operation, and complications and added costs that may result from such a transfer.
- *Repatriation of profit*: Foreign companies prefer having full freedom to remove any profits from the country, whereas the government may seek to impose limitations that would require companies to invest or keep profits within the country. The tendency in recent years has been to move away from such limitations.
- *Local content and training*: Many nations that have oil and gas production wish to increase the use of local labor and services in the petroleum industry, and to increase the capabilities of local firms and personnel as participants in the industry. This may be reflected in provisions for oil companies to give certain preferences to local employment and procurements, and to support the development of skills locally.

12.9 The structure of fiscal terms: balancing risks and reward

Consider the following problem: some countries are known to require a large signature bonus—many million dollars—on signing production sharing contracts. This secures substantial and immediate revenue to the country's treasury. Can there also be detriments to requesting such bonuses, which in the long term may more than offset the short-term benefit?

12.9.1 A company perspective on signature bonuses

It is useful to consider the issue first from the oil companies' perspective. There is generally a large economic risk associated with exploration activities: seismic acquisition, interpretation work, and especially drilling are costly undertakings, and the prior probability of making a viable discovery on a particular drilling prospect is usually no more than 20%. Oil companies therefore face the prospect of investing large sums on exploration, perhaps in the order of \$50–100 million, for a 10–20% probability of making a valuable discovery.

If the company is to accept such a costly and risky exploration venture, then the possible reward must be very large, that is, there must be prospects for the oil company to obtain a very large value from a discovery if it is successful. Not only must the possible discovery be valuable in terms of reserves, the company must also be in a position to expect to keep a large portion of that value for itself, rather than transferring it to the government.

A signature bonus in effect adds to the costs of exploration. From the company's perspective, due to the economic risks involved, this added up-front cost must be offset by a much larger reduction in later fiscal burdens, compared to what would have been acceptable without the signature bonus. This relationship is determined not only by the risk of not finding a valuable discovery, but also by the tendency for the companies to put an economic premium on risk. This premium on risk is normally reflected in the discount rates applied in calculations of net present value (NPV) and expected monetary value (EMV) for investment decision purposes; a high discount rate (including a large risk premium) has the effect that the payment of a certain sum early in the project will have a much larger economic significance than the payment of the same sum later in the project. Oil companies are thus prepared to pay large signature bonuses only if the fiscal burden on any production will be more lenient than they otherwise could have been.

12.9.2 A government perspective on signature bonuses

When negotiating for licensing terms, the government faces two basic challenges:

(1) exploit its bargaining position to the fullest extent;
(2) strike the right balance between the different benefits that it can obtain.

In view of the first challenge, getting a large signature bonus is always better than getting a small signature bonus, *if all other things are assumed to be equal.* This is a matter of taking due advantage of a position, of not selling the nation's potential for less than its worth.

In view of the second challenge, all other things should not be assumed to be equal. On the presumption that the government's negotiators will be able to exploit their bargaining position to the fullest extent, there are trade-offs to be made between the different possible benefits. A big signature bonus can be obtained—but only by foregoing other benefits, such as a larger agreed production share or more extensive work obligations. The expected benefit that must be foregone later is likely to be much larger than the signature bonus that can be had up-front. This is due to the risk involved, augmented by the oil companies' tendency to put a time-dependent premium on risk in their economic evaluations.

Does it follow that a large signature bonus is always wrong? Not necessarily. There may be situations in which a government finds reason to put an even larger premium on risk and on up-front money than the companies do. The government might perceive—rightly or wrongly—that it faces an extraordinary situation in which early cash is an urgent necessity, regardless of the possible long-term benefits which are therewith foregone. A large signature bonus might be applied to build a competent petroleum administration within the government so that it will subsequently be in a better position to manage resources for the nation's benefit. Furthermore, negotiators might also recognize the signature bonus to be a relatively convenient parameter to negotiate, with tangible and prompt benefits, compared to the more uncertain and distant effects of other parameters.

Even when recognizing that large signature bonuses can sometimes be useful and justified, there are at least three cautions to be made. First, there are dangers of integrity and lacking transparency associated with large, one-off payments. Second, if a government has urgent and justified financial needs for which it might look to a signature bonus, it can be recommended to first consider regular, international sources of finance. Third, if a large signature bonus can be justified by extraordinary circumstances, one would not expect these circumstances to be repeated many times—that is, next time, the government should be in a position to aim for larger, long-term benefits rather than a big signature bonus.

12.9.3 Risk properties of different fiscal components

The components of the fiscal regime have different consequences for the distribution of risk. For this purpose, the concept of risk can be understood as follows: a fiscal component creates high risk for either party (government or oil firm) if it has the potential to strongly affect the overall economic outcome for that party depending on uncertainties which that party cannot influence. Risk can be assessed from the perspective of the situation immediately after an agreement has been made between the government and oil firm(s) on licensing a particular area for petroleum exploration.

At that point, the project of exploring in that license is inherently risky: the existence and quantity of any petroleum in the licensed area is unknown, as are future oil prices and the cost of developing the reserves. Therefore, the overall value of the license is highly uncertain; it may range from negative (if money will be spent on exploration and nothing is found) to hugely positive (if a large field is found, effectively developed, and then encounters high oil prices).

No fiscal regime will make this risk go away, but the fiscal regime influences the allocation of risk between the state and the oil firms. Risk entails not only the possibility of losing money, but also the possibility of making large amounts of money if a valuable discovery is made.

It is possible to envisage two extreme solutions:

- On one extreme, the fiscal regime consisting only of a signature bonus (which, as a consequence, might be rather large). In this case, the entire risk associated with resources in the license is left with the oil firm. The government has no risk; it receives the signature bonus and nothing more from the license even if a large discovery is made (however, the government might benefit indirectly, as the development and operations of the field might create more economic activity, resulting in more tax revenues).
- On the other extreme, an arrangement by which all revenues accrue directly to the government. Commercial firms may be contracted to carry out exploration, development, and operational tasks, for which they are compensated independently of discoveries and production. In this case, the entire risk associated with resources in the license resides with the government.

The first extreme solution (company risk only) might have been encountered in the early years of international oil exploration. The second extreme solution (government risk only) resembles the actual situation in countries where a state-owned firm has exclusive rights to all petroleum production (Saudi Arabia, Mexico, and others, on the presumption that their national oil firms are considered as extensions of the state). In other countries, various fiscal components are combined so as to distribute risk in ways that fall between the two extremes.

Following are comments on how the different fiscal components affect the distribution of risk:

Signature bonus: The entire risk is left with the oil firm, and the firm's risk is further augmented by having to pay money to the government which will not be recovered even in the event of a non-discovery. The signature bonus provides the government with an assured up-front payment, but no share of future production value. For the oil firm, it is an up-front payment that will not be recovered even if no petroleum is found.

Area rental fee: The entire risk is left with the oil firm. The risk properties of an area rental fee resemble those of a signature bonus, but somewhat moderated by the possibility that the oil firm may relinquish (parts of) the area and thus reduce the area rental fee, as may be allowed under the license terms.

Royalty: Risk is shared between government and oil firm. Royalty is usually levied as a percentage share of gross production. Government will receive income only if production takes place, dependent on the amount of production and its value (oil prices), but independent of the cost of production. For the oil firm, the risk from royalty is clearly less than from a signature bonus of comparable value, since there will be no royalty in the event of a non-discovery. However, the oil company faces a risk of having to pay royalty on production from a field that has weak economics, especially if a development decision was made on assumptions that later turn out to have been too optimistic.

Production sharing—scale dependent on production rates: Government carries large share of risk. An example is shown in Table 12.3. Compared to royalty, production sharing transfers more risk from the contracting oil company to the state. As a consequence, oil companies generally find large production shares for the state easier to accept than correspondingly large royalty shares. In the event of a very valuable discovery, the state stands to gain more from a production sharing clause than would have been possible from royalty only.

Production sharing—scale dependent on R-factor: Government carries large share of risk. In some production sharing contracts, the state will be entitled to larger shares of production if total project economics will exceed certain defined levels of profitability. The R-factor provides a fairly good indicator of overall project economics, including the effect of high or low oil prices. This therefore entails an even larger transfer of risk from the oil firms to the state.

The five fiscal components mentioned above can thus be ranked in terms of how they affect the distribution of risk between the state and the oil firms:

1. signature bonus (*lowest risk for the state, highest risk for company*);
2. area rental fee;
3. royalty;
4. production sharing, scale dependent on production rates;
5. production sharing, scale dependent on R-factor (*highest risk for the state, lowest risk for company*).

Following are two additional fiscal components which cannot be uniquely ranked in the above list, but which would tend to be at or near the bottom of the list:

Corporate income tax: Government carries large share of risk. Corporate income taxation, as found in many countries and applied businesses in general, is based on total revenues minus total costs. In terms of risk distribution, it differs from production sharing in two important respects. First, corporate income taxation does not have a scale of different percentages dependent on production or product return, but rather a flat tax rate. This tends to transfer less risk to the government than is the case with production sharing. Second, corporate income taxation usually applies to the totality of one firm's business in a country, whereas production sharing applies to individual projects. This tends to transfer more risk to the government.

Special petroleum tax: There are numerous possible ways of designing a special tax for petroleum activities. The presumption here is of a tax intended to capture resource rent deriving from the extraction of limited natural resources, which makes petroleum a different type of business than ordinary manufacturing and trade. A special petroleum tax is particularly relevant in concessionary systems, whereas in contractual systems this purpose can be served by the production sharing provision. The Norwegian special petroleum tax is a 50% tax on income from petroleum production, additional to the 28% regular corporate tax, but with an extra deduction related to past capital investments. The extra deduction (and certain other provisions) tends to transfer more risk from the companies to the government than is the case with ordinary corporate income tax, which helps explain why companies will still invest in Norwegian petroleum in spite of the combined 78% tax on most profits.

12.9.4 The possible benefits of a high-risk strategy

Each country that has petroleum prospects and wishes to involve commercial enterprises in the process of exploration and development needs to establish a fiscal regime suited to its situation and needs. It was argued above that if a country has urgent financial requirements, this may justify an approach emphasizing early cash and low risk, even if this is likely to forego much larger (but also riskier) revenues later.

Norway is an example of a country that did not take a low-risk, early cash approach. Due in large part to a strong shipping industry, Norway was in the privileged position of being a reasonably developed country with orderly national finances when the possibility of oil discoveries was first recognized. An offer made by an American firm in the 1960s to acquire all offshore petroleum rights for US$ 1 million was easily turned down. Instead of leaving all risk and all upside potential with commercial firms, successive governments have gone out of their way to take on financial risks which otherwise would have been borne by the commercial firms. This strategy has become more pronounced as the nation's financial position grew stronger from accumulated oil revenues (Al-Kasim, 2006).

The flip side of taking on more financial risk is that it puts the state in a position to make even larger gains when petroleum resources turn out to be of high value. In terms of financial wealth accumulation, the Norwegian high-risk strategy has achieved results to an extent that would not have been possible by emphasizing early cash and low risk.

Luck, however, also plays a part: in the 1960s, oil prices were at $3 per barrel. Had they stayed at that level, as was widely expected, the $1 million offer might have been assessed more favorably in retrospect.

Political risk is another factor which tends to favor a fiscal regime that places a large degree of financial risk and potential reward with the government. The economic rationale for a low-risk, early cash approach (i.e., large signature bonus) is that oil firms will accept making a risk-exposed payment early against the promise of extracting large rewards later if valuable resources are found. The problem then is that by the time the valuable resources are produced, citizens and politicians of the producing country may become disenchanted with the large amounts of money being made on their resources by foreign firms, while recalling little or no tangible benefit from the initial signature bonus. This has tended to trigger political changes leading to nationalization, or otherwise appropriating value from the oil firms beyond what had been agreed. Such developments took place in a number of countries in the 1970s, and again in the new millennium (Bolivia, Venezuela, and Russia).

In contrast, a properly drafted fiscal regime in which the state assumes a large portion of risk can be more robust and provide greater confidence in the face of possible change. The Norwegian fiscal system has undergone certain adjustments, but no far-reaching change since the 1980s (to 2008). It has provided a fiscal framework that has been considered broadly acceptable to the industry as well as political majorities under a wide range of market conditions which have occurred over that period. This degree of regulatory stability is valuable in itself, because lack of political and regulatory predictability tends to discourage investment in a country. A comprehensive review of the economics of production sharing contracts is provided by Bindemann (1999).

12.10 Summary

The licensing regime is the system applicable in a country under which oil firms may acquire legal rights to explore for and produce petroleum. The licensing regime forms basis for the fiscal regime, which is the set of rules of law and/or contract for allocating the economic benefit and costs of petroleum activities between the state and commercial firms. Concessionary and contractual systems are main forms of licensing regimes. In concessionary systems, the state grants the rights to commercial firms to explore for, produce, and own petroleum in a designated area, subject to the laws of the country and with the obligation to pay taxes and other benefits to the state. Contractual systems are based on the premise that commercial firms undertake to find and produce petroleum for the benefit of the state, for which the firms will have certain rights to compensation as stipulated in the contract. Production sharing contracts and service contracts are main variants under contractual systems.

Provisions for cost recovery and for assigning value to petroleum that is produced, as well as for dealing with natural gas, are main issues that concessionary as well as contractual systems must address. Depending on agreed terms, the state may be entitled to receive a variety of financial benefits, such as signature bonus, royalties, profit shares, taxes, etc., each of which has different implications for risk sharing and incentives for the firms and the state.

References

Al-Kasim, F. The Oxford Institute of Energy Studies, Oxford, England. PennWell Publishing Co, Tulsa, Oklahoma, USA. 2006. *Managing Petroleum Resources; The Norwegian Model in a Broad Perspective*. Oxford Institute of Energy Studies.

Bindemann, K. 1999. *Production Sharing Contracts: An Economic Analysis*. Oxford Institute of Energy Studies.

Johnston, D. 1994. *International Petroleum Fiscal Regimes and Production Sharing Contracts*. Pennwell Publishing Company.

The economics of petroleum operations and investments

People and firms usually engage in business primarily to make money. Firms need procedures for determining the economics of what they do. This chapter introduces two important aspects of how companies manage their economics: financial accounting and project economics analysis. As support for the latter, cost estimating and probability distribution are introduced.

Financial accounting is the system for recording and reporting all transactions that compose a firm's financial situation. Many managers and professionals in a petroleum company come across this system almost daily when entering approvals and cost codes on expense items, following up on budgets, etc. Firms have accounting departments which run the financial accounting system. The products of this system are the company's *financial statements*, which communicate the firm's financial achievements to its owners and other interested parties.

Project economics analysis is a tool for supporting decisions on investments. It is based on the presumption that an investment is attractive if the net value of expected future payments generated by it is positive. Project economics analysis always deals with payments and values which are expected to occur in the future, and which usually cannot be precisely known today. It depends on estimating the costs of a project, and a structured approach to analyzing the uncertainties of values in the future.

The sections below first introduce financial statements, and then basic accounting procedures leading to those statements. Further sections introduce cost estimation, project economics analysis, and probability distribution.

13.1 The financial statements of companies

Many petroleum firms are owned by many shareholders and have their shares traded on public exchanges, such as the New York Stock Exchange or London Stock Exchange. Such companies are sometimes referred to as *public companies*, that is, companies that the general public can invest in as shareholders. The firm is then required to publish comprehensive financial statements. The most important financial statements are found in the firm's annual report. This is a document of usually more than 100 pages that can be downloaded from the company's website, or obtained as a paper booklet. For anyone who may need to understand a company's financial position and what the company does, the annual report is a logical place to start. In this section, parts of the annual report of BP for 2007 will be used as an example. It can be downloaded from the firm's website, www.bp.com.

13.1.1 Contents of the annual report

The annual report of a public company contains a wealth of information, much of which must follow prescribed norms determined by regulatory authorities and accounting standards. Important parts of the annual report for a petroleum firm include the following, with some variations:

- *Board of Directors' report*: A text statement from the Board reporting on important developments of the firm.

- *Financial statements*: The key statements are the income statement and balance sheet, which are further explained below.
- *Notes to the financial statements*: Explaining accounting principles and providing further details.
- *Oil and gas reserves statement*: Detailing quantities of oil and gas that the firm will be entitled to produce in future years (further explained below).
- *Auditor's report*: A statement by an audit firm based on verification of the financial statements.

In addition, annual reports usually contain information that is not necessarily required by regulations, such as highlights and images of the firm's operations.

13.1.2 The income statement

Table 13.1 shows the group income statement for BP, found on p. 96 of BP's *Annual Report and Accounts 2007*.

Table 13.1: BP's group income statement for 2007.

Group Income Statement for the year ended 31 December				$ million
	Note	**2007**	2006	2005
Sales and other operating revenues	5	**284 365**	265 906	239 792
Earnings from jointly controlled entities – after interest and tax		**3 135**	3 553	3 083
Earnings from associates – after interest and tax		**697**	442	460
Interest and other revenues	6	**754**	701	613
Total revenues		**288 951**	270 602	243 948
Gains on sale of businesses and fixed assets	7	**2 487**	3 714	1 538
Total revenues and other income		**291 438**	274 316	245 486
Purchases		**200 766**	187 183	163 026
Production and manufacturing expenses		**25 915**	23 793	21 092
Production and similar taxes	8	**4 013**	3 621	3 010
Depreciation, depletion, and amortization	9	**10 579**	9 128	8 771
Impairment and losses on sale of businesses and fixed assets	10	**1 679**	549	468
Exploration expense	16	**756**	1 045	684
Distribution and administration expenses	12	**15 371**	14 447	13 706
Fair value (gain) loss on embedded derivatives	34	**7**	(608)	2 047
Profit before interest and taxation from continuing operations		**32 352**	35 158	32 682
Finance costs	18	**1 110**	718	616
Other finance (income) expense	19	**(369)**	(202)	145
Profit before taxation from continuing operations		**31 611**	34 642	31 921
Taxation	20	**10 442**	12 331	9 473
Profit from continuing operations		**21 169**	22 311	22 448
Profit (loss) from Innovene operations	3	**–**	(25)	184
Profit for the year		**21 169**	22 286	22 632
Attributable to				
BP shareholders		**20 845**	22 000	22 341
Minority interest		**324**	286	291
		21 169	22 286	22 632

The income statement describes the result of the firm during a particular period, which is here the calendar year of 2007. In addition, columns for 2006 and 2005 are included for reference. Most firms, like BP, have financial reporting years coinciding with the calendar year, but some firms report for 12-month periods ending at a different time, which would be indicated on top of the statement.

The number that probably attracts most interest is the profit for the year, which for 2007 was US$ 21,169 million (21 billion). This is the "bottom line" (though not quite on the bottom of the page) of how much money the entire BP Group made during that year, after having provided for all expenses and taxes.

Profits can be described simply as income minus costs, or in the slightly different terminology used here: revenues minus expenses. The total revenues for 2007 can be identified as US$ 291,438 million. Most of this comes from selling oil, gas, and other products by BP itself (284,365), but there are also some revenues from entities where BP has a lesser interest, and from selling large assets at a profit.

Below the revenues is a specification of the various expenses that must be deducted in order to arrive at the profit. The biggest expense item is purchases. This item is particularly large because BP engages extensively in oil trade, where cargoes of oil are bought and resold as a way of increasing profits. The production and manufacturing expenses and the distribution and administration expenses are the items most closely associated with running BP's operations, including BP's share of operations where it is a non-operating partner.

The item "Depreciation, depletion, and amortization" deals with the cost of long-term assets, as will be further explained below.

As a legal structure, BP is not one company, but many companies, including affiliates in a large number of countries. The group has a parent company, BP p.l.c., which ultimately owns the other companies in the group. The income statement shown above is a *consolidated* statement, that is, it covers all of BP including the parent company and all affiliates. It is therefore referred to as the group income statement. To arrive at the consolidated financial statements, similar statements must be prepared for every company in the group. Some companies are counted as part of the BP Group even though other parties outside BP also have some shares in them, usually a minority share. As a result, the profit of the entire BP Group contains a small portion that is earned by those outside parties, and not by BP itself. This is specified on the three last rows in Table 13.1.

The income statement also makes a distinction between continuing operations and operations which are no longer part of BP at the end of the reporting year. This concerns mainly a petrochemical operation, called Innovene, which BP sold in 2006.

13.1.3 The balance sheet

Table 13.2 shows BP's group balance sheet, which is found on p. 97 of BP's *Annual Report and Accounts 2007.*

The balance sheet describes the financial situation of the company at one particular point in time, which is here at the end of 2007 (and end of 2006 included as an additional column). This is an important difference between the two main financial statements: the income statement covers events during a period (all of 2007), whereas the balance sheet covers the situation at one point in time (December 31, 2007).

Figure 13.1 outlines the structure of the balance sheet. It has two sides: on one side, the *assets*, which is everything that the firm owns. On the other side, the *liabilities* and *equity*, which show how the firm has acquired financial resources enabling it to own the assets.

Table 13.2: BP's group balance sheet for 2007.

Group Balance Sheet at 31 December			$ million
	Note	**2007**	2006
Non-current assets			
Property, plant, and equipment	23	**97 989**	90 999
Goodwill	24	**11 006**	10 780
Intangible assets	25	**6 652**	5 246
Investments in jointly controlled entities	26	**18 113**	15 074
Investments in associates	27	**4 579**	5 975
Other investments	29	**1 830**	1 697
Fixed assets		**140 169**	129 771
Loans		**999**	817
Other receivables	31	**968**	862
Derivative financial instruments	34	**3 741**	3 025
Prepayments		**1 083**	1 034
Defined benefit pension plan surplus	38	**8 914**	6 753
		155 874	142 262
Current assets			
Loans		**165**	141
Inventories	30	**26 554**	18 915
Trade and other receivables	31	**38 020**	38 692
Derivative financial instruments	34	**6 321**	10 373
Prepayments		**3 589**	3 006
Current tax receivable		**705**	544
Cash and cash equivalents	32	**3 562**	2 590
		78 916	74 261
Assets classified as held for sale	3	**1 286**	1 078
		80 202	75 339
Total assets		**236 076**	217 601
Current liabilities			
Trade and other payables	33	**43 152**	42 236
Derivative financial instruments	34	**6 405**	9 424
Accruals		**6 640**	6 147
Finance debt	35	**15 394**	12 924
Current tax payable		**3 282**	2 635
Provisions	37	**2 195**	1 932
		77 068	75 298
Liabilities directly associated with the assets classified as held for sale	3	**163**	54
		77 231	75 352
Non-current liabilities			
Other payables	33	**1 251**	1 430
Derivative financial instruments	34	**5 002**	4 203
Accruals		**959**	961
Finance debt	35	**15 651**	11 086
Deferred tax liabilities	20	**19 215**	18 116
Provisions	37	**12 900**	11 712
Defined benefit pension plan and other postretirement benefit plan deficits	38	**9 215**	9 276
		64 193	56 784

Table 13.2 (*Continued*)

Total liabilities		**141 424**	132 136
Net assets[a]		**94 652**	85 465
Equity			
Share capital	39	**5 237**	5 385
Reserves		**88 453**	79 239
BP shareholders' equity	40	**93 690**	84 624
Minority interest	40	**962**	841
Total equity	40	**94 652**	85 465
Total liabilities and equity[b]		**236 076**	217 601

[a]Net assets = Total assets – Total liabilities = Total equity.
[b]The item "Total liabilities and equity" is not shown on BP's statement, but is included here for reference.

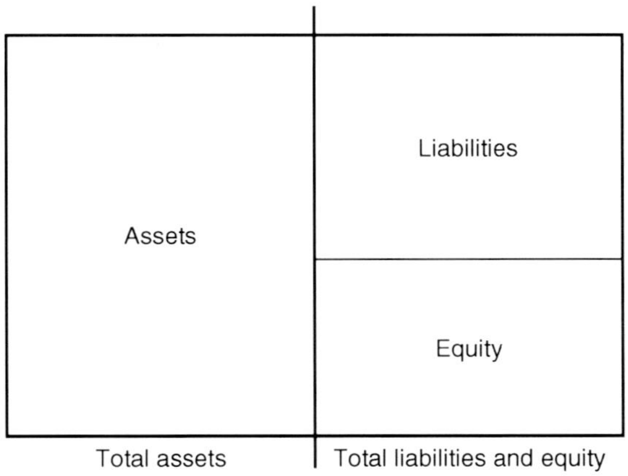

Figure 13.1: Outline structure of a balance sheet.

The two sides are often shown on separate pages, although in BP's case they appear on one page with the liabilities and equity below the assets.

The structure outlined in Figure 13.1 can be recognized in BP's balance sheet (Table 13.2). The total assets were US$ 236,076 million at the end of 2007. This is also the sum of the liabilities and equity, since equity per definition is the value of assets minus liabilities. (It has been customary to specify the total liabilities and equity on the balance sheet, although it is always the same as the total assets.)

Assets consist of *fixed assets*, other *non-current* (long-term) assets such as loans to other parties, and *current assets*. Fixed assets include the value of tangible installations owned by BP, and also assets in which the firm has indirect control. *Goodwill* relates mainly to the price that BP has paid for the acquisition of other companies in the past.

Current assets are those which are normally exchanged frequently in the course of business. They include bank deposits, receivables from customers, prepaid expenses, product in inventory, and short-term investments.

A similar distinction is made between *current* and *non-current liabilities*. Current liabilities are debts and payables that are settled within a short period of time, usually within a year. The cost of goods which have been received by the company but not yet paid on the balance date is a typical current liability. Long-term loans taken by the company, and pension obligations, are typical non-current liabilities.

The *equity* of the company is the value that can be attributed to the shareholders. Some what simplified, the equity comes from two sources: share capital originally paid in by shareholders some time in the past and accumulated profits over the years that have not been repaid to shareholders. Profitable companies usually pay dividends to the shareholders, which is a way of rewarding shareholders for having invested in the profitable company. Dividends are not accounted as an expense; rather, they are a way of utilizing the profit.

13.1.4 Market value and book value

The equity as shown on the balance sheet is a measure of the net value of a company, that is, its book value. There is also another measure available of a company's worth: the company's market value based on its share price on the stock exchange. Shares which are traded on an exchange have a market price which may change by the minute, reflecting the price at which investors are buying and selling shares. Multiplying the share price with the number of shares gives the market value of the company.

Since share prices fluctuate all the time, so does the market value of a company. These fluctuations are often influenced by events that are entirely beyond the firm's control, such as oil price movements and expectations for the global economy. However, one indication that a company is perceived as successful and well managed is that it has a market price that is clearly higher than its equity book value.

13.1.5 Accounting for installations: fixed assets and depreciation

Timing is an important issue in financial reporting. Accountants put great emphasis on not only recording all incomes and costs, but also assigning them to the correct period of time, so as not to distort the earnings of the firm.

Investments in installations that the company will use for many years are an important timing issue in accounting. If a manufacturing company in the year 2010 pays \$5 million for a machine that will operate for the next 10 years, how will this be accounted for? This is done as depreciation, as follows (assuming the machine is paid for at the end of 2010 and starts operating at the beginning for 2011):

- In 2010, the machine is accounted for as a fixed asset. In Table 13.2 it would be included under "Property, plant, and equipment at cost," that is, \$5 million. There is no direct impact on the income statement for 2010.
- The machine will be depreciated over its lifetime. Depreciation is an annual cost reflecting a reduction in value of the asset. If the machine is assumed to last for 10 years, has no residual value, and performs basically the same function for those 10 years, then its linear depreciation over 10 years will be appropriate. This implies that the value of the machine is reduced by \$0.5 million each year on the balance sheet, while a cost of \$0.5 million is included as depreciation on the income statement.
- The balance sheet will therefore include the value of the machine as \$4.5 million for 2011, \$4 million for 2012, and so on, down to 0 for 2020. If the machine continues to operate after 2020, there will be no further depreciation or asset value included for that time.
- The machine will hopefully generate income for the company. The revenues will be included as part of the income statement, as will also the costs of running the machine.

Correct accounting for fixed assets requires that expenditures are properly classified as either expenses or capital cost items. A machine for $5 million would clearly be a capital cost item, and accounted for as an asset as described above. This is due to a combination of long life and large cost. In principle, equipment items having a useful life beyond the year of purchase should be classified as capital assets and depreciated. This however requires more accounting work than simply expensing the item, and in practice, small individual items usually get expensed. A pencil sharpener costing $1 may also last for 10 years, but one need not take the trouble of accounting for this as an asset with an annual depreciation schedule. Companies as well as national accounting standards have guidelines in this regard.

The simple example above features linear depreciation over 10 years, that is, the same amount of $0.5 million in each year. For oilfield installations, this would usually not be an appropriate depreciation schedule. Oilfields tend to have a peaked production curve: it quickly reaches a plateau of high production, which is held for a few years, before entering into a long decline. Depreciation should then not be the same amount each year, but is prorated to production. This is called *unit-of-production depreciation.*

Simply put, the development cost of the field is related to the estimated recoverable reserves. If a field initially has 50 million barrels of proven recoverable reserves, and is developed at a cost of $200 million, then depreciation will be $4 for each barrel produced ($200 million/50 million barrels = $4 per barrel). Consequently, if a field during a particular year produces 15% of its original proven reserves, then depreciation during that year will be 15% of development costs. Adjustments may have to be made later if the reserve estimate changes or additional development costs are incurred.

13.1.6 Accounting for oil and gas reserves

Resulting from successful exploration work, petroleum firms have discovered or acquired reserves of oil and gas in the ground which they can produce as sources of income. The values of the reserves are not accounted for as part of the balance sheet, where the firm's ordinary assets are found.

One reason for this is that in most countries the firms do not own the reserves as long as they are in the ground; they are then the property of the nation in whose jurisdiction they are located. Firms may have acquired certain rights to produce the reserves and keep part of the proceeds, under either concessionary systems or contractual systems as reviewed in Chapter 12.

However, the reserves are reflected in the annual report in two ways. Even if the values of reserves are not in the balance sheet, the costs of finding them are there. To the extent exploration activities have been successful based on certain criteria, the exploration expenditures can be capitalized. In BP's balance sheet (Table 13.2), those capitalized exploration expenditures are part of the *intangible assets*.

Oil firms' annual reports also contain information on oil and gas reserves as supplementary information. Table 13.3 shows one of several tables in BP's 2007 annual report that provide such information. It details where BP has oil reserves, whether they are held by BP itself or through a firm partly controlled by BP (mainly TNK-BP in Russia), and how reserves changed during the year.

The information on firms' oil reserves has been subject to particular interest in recent years (to 2008) because many oil firms based in Europe and North America have found it difficult to fully replace the oil they produce with new reserves, which in the long term may erode their position as petroleum producers.

Table 13.3: Supplementary information on oil reserves from BP's 2007 annual report.

Crude oil									million barrels
	UK	Rest of Europe	US	Rest of Americas	Asia Pacific	Africa	Russia	Other	Total
Subsidiaries									
At 1 January 2007									
Developed	458	189	1 916	130	67	193	–	88	**3 041**
Undeveloped	146	97	1 292	237	86	512	–	482	**2 852**
	604	286	3 208	367	153	705	–	570	**5 893**
Changes attributable to									
Revisions of previous estimates	(1)	(25)	18	(29)	(7)	(133)	–	(27)	**(204)**
Purchases of reserves-in-place	–	–	25	–	–	–	–	8	**33**
Discoveries and extensions	–	31	60	1	2	93	–	–	**187**
Improved recovery	7	1	99	6	5	12	–	1	**131**
Production[b]	(73)	(19)	(169)	(27)	(15)	(71)	–	(80)	**(454)**
Sales of reserves-in-place	–	–	(94)	–	–	–	–	–	**(94)**
	(67)	(12)	(61)	(49)	(15)	(99)	–	(98)	**(401)**
At 31 December 2007									
Developed	414	105	1 882	115	61	256	–	104	**2 937**
Undeveloped	123	169	1 265	203	77	350	–	368	**2 555**
	537	274	3 147	318	138	606	–	472	**5 492**
Equity-accounted entities (BP share)									
At 1 January 2007									
Developed	–	–	–	221	1	–	2 200	520	**2 942**
Undeveloped	–	–	–	139	–	–	644	163	**946**
	–	–	–	360	1	–	2 844	683	**3 888**
Changes attributable to									
Revisions of previous estimates	–	–	–	178	–	–	413	167	**758**
Purchases of reserves-in-place	–	–	–	–	–	–	16	–	**16**
Discoveries and extensions	–	–	–	2	–	–	283	–	**285**
Improved recovery	–	–	–	59	–	–	–	1	**60**
Production	–	–	–	(28)	–	–	(304)	(73)	**(405)**
Sales of reserves-in-place	–	–	–	–	–	–	(21)	–	**(21)**
	–	–	–	211	–	–	387	95	**693**
At 31 December 2007									
Developed	–	–	–	328	1	–	2 094	573	**2 996**
Undeveloped	–	–	–	243	–	–	1 137	205	**1 585**
	–	–	–	571	1	–	3 231	778	**4 581**

13.2 Accounting routines

Managers or professionals who are authorized to incur costs on behalf of the firm usually also come in contact with accounting routines. These routines serve not only the financial reporting system to generate the annual report, but also a number of other purposes such as project management, general managerial control, tax reporting, recovery of costs from partners, recovery of costs under production sharing contracts, insurance issues, etc., as the case may be.

13.2.1 Recording transactions

An *account* is a category of value, designated by an account number (often four digits) and an account name. There are essentially four types of accounts: income, expense, assets, and liabilities (including equity). The income and expense accounts are added up to make the income statement (Table 13.1). The assets and liabilities accounts are added up to make the balance sheet (Table 13.2).

Accounting systems are based on recording *transactions*. A transaction is an event of economic consequence. Transactions should have appropriate documentation such as invoices, time sheets, bank statements, etc.

There are always two sides to a transaction: value is obtained from something, and is applied to something else. Accountants refer to the two sides of transactions as *debit* and *credit*. Put simply, debit signifies *to*, and credit signifies *from*. The two sides must add up to the same amount. An accounting transaction therefore describes an amount being obtained from something (*credit*) and applied to something else (*debit*). The "somethings" are identified as accounts.

Table 13.4 shows some examples of such transactions.

13.2.2 Accounting for costs: what, why, and who

To effectively manage an organization, it may be necessary to know *what* money has been spent on, *why* it was spent, and *who* is responsible. Cost items then need to be classified to make such knowledge possible.

Debiting costs to an expense account determines the "what" question: it shows what sorts of items money has been spent on, such as spare parts, contractor work, fuels, travel expenses, salaries, interests, taxes, etc. This information is needed to generate the financial statements. Companies have internally defined charts of accounts that they use to classify costs, although there are also certain standard charts of accounts that can be applied directly or modified for internal use.

The "why" question relates to projects or other types of activities. Spending money on some item is not an end to itself; it is done for a purpose. The purpose may be a project that has been established as such. Many organizations therefore have a system of codes, additional to the account codes, to classify costs by projects. Costs that are not part of any particular project may instead be classified by general activity codes.

The "who" question relates to responsibilities for costs. Organizations have a number of managers who may be authorized to incur costs for the firm, and who should be held responsible for those costs relative to a budget. The activities under the responsibility of a manager can be designated as a *cost center*. Costs are classified by cost center codes, which make it possible to produce departmental cost reports that can be compared to a budget.

Table 13.4: Examples of accounting transactions.

Event	Accounting	Accounts, generalized
We delivered crude oil with an agreed sales value of $5,123,456 to a buyer	Debit 5,123,456	Trade receivables (asset)
	Credit 5,123,456	Sales revenues (income)
We received payment of $5,123,456 for crude oil to our bank account	Debit 5,123,456	Bank deposit (asset)
	Credit 5,123,456	Trade receivables (asset)
We received spare parts to our warehouse; cost $54,321	Debit 54,321	Inventories (asset)
	Credit 54,321	Trade payables (liability)
We paid supplier for spare parts based on invoice	Debit 54,321	Trade payables (liability)
	Credit 54,321	Bank deposit (asset)
We used the spare parts worth $54,321 as a replacement at our installation	Debit 54,321	Spare parts (expense)
	Credit 54,321	Inventories (asset)

13.3 Project cost estimation

Estimating the cost of a project or activity that may be undertaken is a requirement for making an informed decision on whether or not to carry it out. It is also a difficult area where gross errors of estimation have sometimes been made. Whereas accounting, as described above, is a task of precision in dealing with events that have occurred, cost estimation is an attempt to assess what will be in the future. This section introduces cost estimating of investment projects and operations.

13.3.1 Purposes and challenges of cost estimation

Costs in petroleum projects, beyond the exploration phase, are usually categorized as *Capex* (capital expenditures) and *Opex* (operating expenditures). These categories broadly correspond to the accounting distinction made above between expenditures classified as assets and as expenses. Capex includes the cost of constructing and installing facilities and drilling wells. Opex includes the recurring costs of operating and maintaining the installations, transportation tariffs, and administration.

Cost estimates serve as a basis for economic evaluations and to support the decision process for company management, project partners, and authorities during the project approval process. Furthermore, the estimates should form a basis for optimizing the technical concept and commercial opportunities, as well as for budgets, approval of expenditure, project control, and reporting. The estimates are also used for evaluating amendments and change orders which may be initiated during the project execution.

Cost estimation can be seen as an analysis of future expected costs under defined rules and conditions. It is an approximation procedure, which provides answers that have significantly less than 100% probability of being correct. An estimate can also be represented as a probability distribution (explained in a later section). The process has intuitive elements of attempting to predict the final outcome of a future expenditure program, even though not all parameters and conditions concerning the project are known or fully defined.

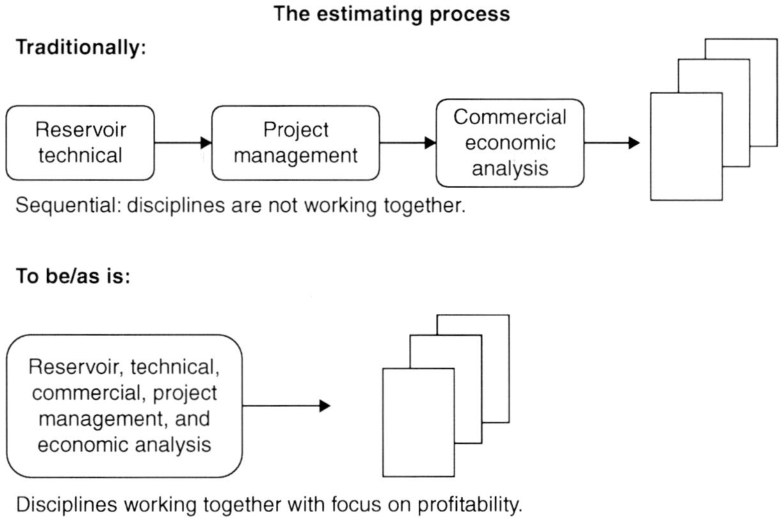

Figure 13.2: The estimating process: sequential and integrated organization of tasks.

The quality or accuracy of an estimate depends on the degree of technical definition of the project, project execution assumptions and definitions, and knowledge of price data for material and services required.

The costs associated with a field development can be related to the different phases as follows:

- exploration costs;
- cost associated with the planning phases;
- cost during the execution of a project;
- drilling costs;
- operating costs;
- maintenance costs;
- costs related to close down and removal of installations.

As in many other areas of work, there are advantages to be obtained from organizing the work in an integrated manner involving close cooperation between disciplines. Traditionally the approach has been a sequential work flow, in which the commercial and economic analyses (including cost estimates) were made after the reservoir and technical functions had worked out their contributions to the project. The difference between the two approaches is illustrated in Figure 13.2.

13.3.2 Synthetic and analytic techniques of cost estimation

Synthetic estimation is characterized by the following features:

- poorly defined technical and operational concepts;
- use of relational or factor-estimating methods;
- use of functional data, capacities for the project, experience, and reference data;
- indirect calculation of the costs based on characteristic data, for example, the total cost for a platform is found by determining the weight of the support structure and the topsides.

Analytic estimation is, on the other hand, characterized by the following features:

- technical and operational concepts are known, the scope is defined, and manpower and complexity can be calculated;
- resources are costed directly based on equipment lists and material take-off;
- an estimate can consist of 200–1000 items.

13.3.3 Synthetic cost estimation

Synthetic and analytic techniques are two different methodological approaches to cost estimation. They can be suited to different circumstances depending on the project phase, maturity of the technical concept, the quality of experience data, and the time available.

There are two variations of synthetic cost estimation: synthetic relational estimates and synthetic factor estimates.

Synthetic relational estimation is less time consuming and often used for project evaluations. It requires little technical documentation, and can therefore be used when technical and operational concepts are poorly defined. Cost is calculated by the following type of function:

$$\text{cost} = \text{Unit cost} \times \text{Units}$$

A synthetic relational estimate can also be used to compare and make estimates for similar facilities with different capacities. The type of equation applied is then:

$$K_1 = K_0 \frac{Q_1}{Q_0} N$$

where K_0 and K_1 are the costs for old and new facilities, Q_0 and Q_1 the capacities for old and new facilities, and N a factor based on experience data, which varies for different types of projects, considering location, year of construction, improvements, etc.

Synthetic factor estimation is more detailed than relational estimation, and mainly used for onshore plants. The cost for the main equipment items of the project must be known and is the basis for estimating overall costs. The factors used are dependent on the type of installation, and factors for process plants are usually between 4 and 10. The complexity of the factor estimate may vary considerably, and can be as simple as the following:

Equipment = 1.
Bulk = 0.97.
Labor = 1.97.

Based on the equipment cost, the total costs are then calculated:

$$\text{Total costs} = (1 + 0.97 + 1.97) \times \text{equipment}$$

13.3.4 Analytic cost estimation

In analytic estimation, all cost items in the project are estimated directly. The method can only be used in the later phases of a project when detailed technical information is available. The method requires more work than the methods mentioned before. The technical concept must be broken down into components that can be estimated. All equipment and bulk goods must be identified and estimated. This requires an equipment list and details for each discipline, such as: piping, steel structures, electro, instrument, etc. Vendor information, data from similar projects, market evaluations, and experience data are used as sources. There is a danger of missing components of required costs in analytic estimates, and a properly structured system is essential. A particular concern is piping, which often is difficult to fully overview until the late project phases.

The most important data sources for cost estimating are with respect to quantities, specifications and drawings, bulk and equipment lists, estimating manuals, analyzed "as built" and reference data, estimating tools, and technical personnel competence. With respect to unit costs, the following important sources shall be considered: quotations/contract data, market data, operating units, estimating manuals, analyzed "as built" and reference data, estimating tools, and the competence of the commercial and technical personnel.

13.3.5 Uncertainties

A cost estimate is a prediction, and as such always associated with uncertainties. The evaluation of the estimate's accuracy is an important part of the estimating job. The more accuracy is claimed for an estimate, the less likely it is to be correct. For this reason, a confidence level is defined for estimates.

The *accuracy* of a cost estimate is given by a symmetric interval around the expected value of the cost estimate. The interval, given as a ± percentage, is a confidence interval corresponding to a confidence level of, for example, 80%. There should then be a probability of 80% for the final cost to be within the interval.

Contingency is often added to a cost estimate. It is the expected costs for unspecified cost elements. Some companies have standards with fixed percentages of total cost as contingency depending on decision gate (DG)/project phase. Deviation from the fixed contingency addition may be acceptable if justified by an evaluation of remaining uncertainty.

Estimates are based on parameters which have a representative value and a variation range. In uncertainty analysis of cost estimates, the uncertainty associated with each input to the cost estimate is established, and the resulting uncertainty of the total estimate calculated.

Input parameters can be specified with probability distributions. An introduction to probability distributions is given in the final section of this chapter. Assuming a normal or lognormal distribution, the input parameter may be specified by the following three values:

- *Pessimistic value*: The value representing the pessimistic outcome of a parameter. Input to uncertainty analysis is given as the *90/10 value* or P_{90}. For a cost item, this is an estimate of the cost which has a 90% probability of not being exceeded; it must therefore be a higher estimate than what is considered most likely.
- *Most likely value*: The value representing the most likely outcome of a variable. Input to the uncertainty analysis is given as the *base estimate*.
- *Optimistic value*: The value representing the optimistic outcome of a variable. Input to the uncertainty analysis is given as the *10/90 value* or P_{10}.

There are also other possible distribution functions that may be used, as reviewed in the last section of the chapter. Based on such values for a large number of input parameters, a Monte Carlo simulation can be run to obtain a probability distribution for the total cost estimate. Figure 13.3 shows an example of a probability distribution that may result from such

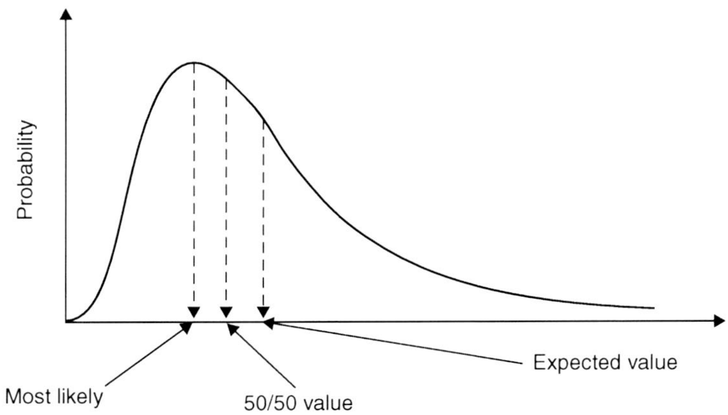

Figure 13.3: A probability distribution and the three measures of central tendency.

a simulation. The horizontal axis defines the possible total costs of the project as outcomes, and the vertical dimension indicates the probability of the various outcomes. The example shows a *skewed* (not symmetric) distribution, which has three different measures of central tendency:

- *Most likely value (mode)*: The single value (or small value interval) that is most likely to occur, that is, the tallest point of the distribution curve.
- *50/50 value (median)*: The value that gives equal probability of underrun and overrun, the point which divides the distribution curve so that the area under the curve is the same on both sides.
- *Expected value (mean)*: The weighed average of all possible outcomes, that is, the point that divides the distribution curve so that the areas of the two sides are in balance.

(These concepts are discussed in the last section of this chapter.)

13.3.6 Cost estimate classes

When planning work on estimation, the appropriate level of accuracy has to be decided, bearing in mind that the work may require significant amounts of analyst work. As a structured approach to this issue, some companies use classification systems, which specify requirements for applying basic assumptions, methods, level of detailing, scheduling, uncertainty analysis, etc., to the estimation process. Table 13.5 shows an example of a cost estimate class system with associated ranges of accuracy.

The estimate classes can be related to DG as introduced in Chapter 8, so that an estimate of a certain class will be required for each DG. Following is an example:

- *DG 0*: Decision to start feasibility studies – cost estimate class *A* required.
- *DG 1*: Decision to start planning – cost estimate class *B* required.
- *DG 2*: Provisional project sanction – cost estimate class *C* required.
- *DG 3*: Project sanction – cost estimate class *D* required.
- *DG 4*: Start operations.

Table 13.5: Example of cost estimate classes, with accuracy ranges.

Class	Description	Accuracy
A	Prospect study	N/A
B	Feasibility study	±40%
C	Concept studies	±30%
D	Pre-engineering	±20%

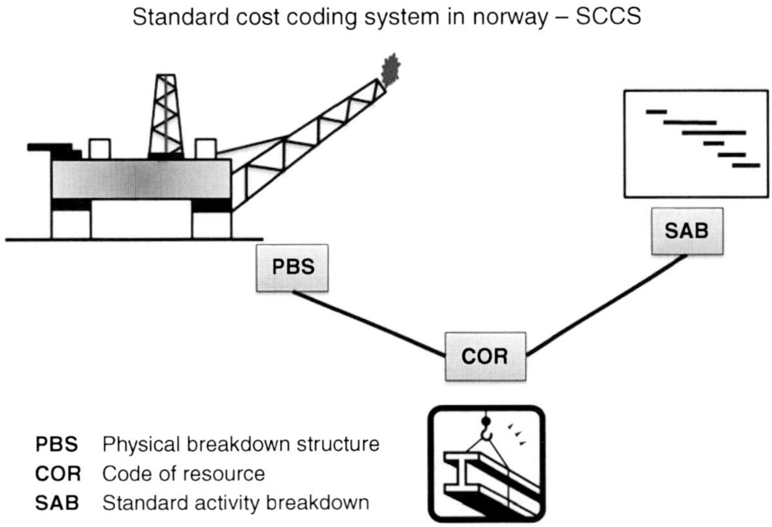

Figure 13.4: Standard cost coding system.

13.3.7 The structure of cost estimates

There are advantages in applying standardized structures for cost estimates. This facilitates preparation, comparisons, and communication of estimates, and the analysis of historical data. A common standard is in use in the Norwegian petroleum sector, called the *standard cost coding system (SCCS)*, and is a part of the Norsok Standard (Norsok, 2005). It consists of three coding structures (Figure 13.4), as illustrated in the following:

- *Physical breakdown structure (PBS)*: Describes the physical composition of the project, such as the main components in an onshore plant, platform, transport system, subsea installations, etc.
- *Code of resource (COR)*: Describes materials and services needed in a project, for example, management, engineering, and equipment.
- *Standard activity breakdown (SAB)*: Describes the breakdown of activities in a project, such as management, procurement, fabrication, and construction.

Each kind of structure is hierarchically broken down. A PBS may be structured as follows:

- A: offshore fields installation:
  - AA: topsides;
  - AB: substructures;

- AC: wells;
- AD: subsea production system;
- ADA: production station.
- B: land-based installation.

The SCCS is applied similarly for different projects. An alternative concept is a work breakdown structure (WBS), which is connected to the contracts required for a project. The ability to use a uniform system such as SCCS has the advantage of facilitating experience data from several previous projects as a basis for future cost estimation.

Using experience data from previous projects requires a database where quantitative information is stored in a structured way. Ongoing projects provide experience data to secure that information is continuously updated. Experience data are used for estimates in the planning phase, checking detailed estimates, and comparisons between projects in the execution phase and with projects run by other operators.

13.3.8 Preparing investment cost estimates (Capex)

The process of developing an investment cost estimate starts with defining the scope of work, framing the project, and defining robust assumptions. Following are important inputs on which the cost estimate should be based:

- the technical design of the project;
- evaluation of markets for required goods and services that need to be evaluated, for instance, anticipated market prices for steel;
- a defined procurement strategy.

Availability of these inputs enables the calculation of expected project costs and a probability distribution as discussed above. Also, an expenditure profile can be calculated, that is, a schedule of when payments will be required, as a basis for financing the project. It is recommended to submit the cost estimates to a separate quality assurance process.

Cost estimates can go wrong, and provide misleading results. Problems may be encountered for reasons such as the following:

- insufficient knowledge by cost analysts of the technical and operating concepts;
- insufficient emphasis on future technology and on project execution philosophies;
- insufficient verification of data;
- use of materials lists that are not updated;
- errors of omission;
- deficient evaluation of markets for required goods and services;
- lack of communication between technical, operations, procurement, and project control personnel.

The uncertainty of cost estimates can be reduced by a high quality of the estimation work. This requires that analysts critically review the individual data as well as the overall structure of the estimate, building on available experience and a good understanding of the project that is to be built. Open channels with the rest of the company are required to ensure that all relevant information is available for the estimation work. Basis and methodology have to be documented using structures thoroughly with focus on close-out formats. Projects that have had cost overruns have typically not been sufficiently defined in the early project phases, or the project management structure has been weak.

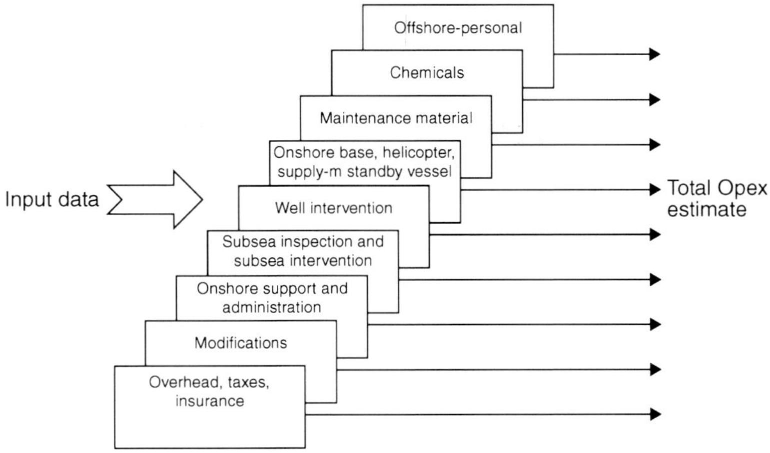

Figure 13.5: The different parts of an Opex estimate.

13.3.9 Preparing operating cost estimates (Opex)

Operational expenditures, Opex, are the ongoing costs for running a petroleum operation. Figure 13.5 shows the different parts of an Opex estimate for an offshore platform-based operation. For an onshore operation, some of the items in the figure can be left out.

Maintenance work is a large component of Opex. For offshore platforms, the maintenance costs are in the order of 50% of the total operation costs. For a refinery, this share is typically 30–35%. Companies may take different approaches to the maintenance of facilities, and it is a function for which particular strategies are required. Petroleum installations tend to have long useful lives, 30 years or more, and may need maintenance programs from the early production phase designed to keep the facility fully functional for a long time. There is also a scope for optimizing total operating costs, that is, the sum of maintenance and other production costs, which can yield large improvements. Some companies have increased maintenance costs in order to reduce production costs, resulting in large overall cost savings. The expected maintenance costs should also be evaluated to identify the optimum balance between planned and unplanned maintenance (prevention vs. repairs).

An operating cost estimate can be established early in a project to serve as management guidance for optimizing the operating phase. The estimate can then be a dynamic document, and should be updated regularly as cost elements become more accurately defined.

13.4 Project economics analysis

Should our organization invest capital in a particular development project? Such decisions in the oil and gas industry sometimes commit very large sums of money, and may shape the future of the organization involved.

This section will introduce the concepts that are commonly applied for analyzing potential projects in preparation for making such investment decisions, large or small. Sections 13.4.1–13.4.4 introduce, by use of a simple example, the key terminology and concepts of project economics, such as cash flows, time value of money, discount rate, present value, and internal rate of return. The remaining parts of this section discuss certain issues which are important in project economics analysis.

13.4.1 Cash flows

Project economics is usually done as *cash flow* analysis. Cash flows are payments, and can normally be identified in retrospect as a movement on the firm's bank statement.

As an example, someone may come up with the following proposal: our organization should spend \$90,000 to buy a machine, which will enable us to increase sales revenues by \$30,000 per year.

Such received information must be checked for completeness. To analyze the project economics of this proposal, the following information is also required:

(1) Is \$90,000 the total expenditure of acquiring the machine, or will there be added initial costs for freight, installation, etc.?
(2) How much will it cost per year to operate and maintain the machine?
(3) For how long can we expect the machine to be useful before it becomes either worn out or obsolete?
(4) Will it have any residual value at the end of that period?
(5) Are the revenues of \$30,000 per year truly additional to the firm's other revenues, or will there be side effects on other parts of the firm's sales?

Table 13.6 shows the cash flows of this project based on the information available. Figure 13.6 shows the same as a histogram, each column representing a year.

For the purpose of illustration, the following answers to the above questions were assumed: (1) additional expenditures for freight, installation, etc., will be \$10,000, so that the total expenditure of acquiring the machine will be \$100,000; (2) annual costs of extra manpower, maintenance, electricity, etc., will be \$10,000; (3) 10 years; (4) no residual value, as we

Table 13.6: Example of a simple cash flow calculation (taxes not considered; unit: US dollars).

Year	Investment	Revenues	Operating	Salvage	Net cash flow
0	−100,000				−100,000
1		30,000	−10,000	0	20,000
2		30,000	−10,000	0	20,000
3		30,000	−10,000	0	20,000
4		30,000	−10,000	0	20,000
5		30,000	−10,000	0	20,000
6		30,000	−10,000	0	20,000
7		30,000	−10,000	0	20,000
8		30,000	−10,000	0	20,000
9		30,000	−10,000	0	20,000
10		30,000	−10,000	0	20,000
11		0	0	0	0

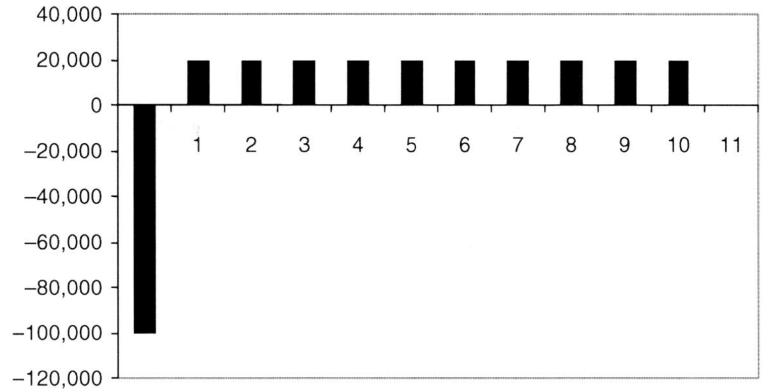

Figure 13.6: Net cash flow from machine project.

assume the machine can be sold for scraps at a value that just covers the cost of removal; (5) no such side effects assumed.

Note the following features of the cash flow table in Table 13.6:

- Each row represents a year. The start year is identified here as year 0.
- Outgoing cash flows are shown as negatives, and ingoing as positives.
- Net cash flow for each year is the net sum of payments (inflows minus outflows).
- As with all data tables, the units should be clearly identified (US dollars in this case).

The project may have an impact on the company's taxes, which is not shown here. The net cash flow shown is therefore a pretax cash flow. With knowledge of the applicable tax system, it would be possible to expand the table with more columns to show taxes and the after-tax net cash flow.

Note that *profit*, and certain other terms found in a firm's income statement, is not cash flow. An income statement contains certain components which are closely linked to cash flows (such as sales revenues and labor costs) and other components which are not cash flows (such as depreciation). The firm's balance sheet also does not show cash flows.

A cash flow for the purpose of calculating project economics can also be a saved expenditure. We might modify the example above so that the machine would not generate any extra revenue, but it would make the organization avoid certain expenditures which otherwise would be incurred, of $30,000 per year. The cash flow would then be precisely the same as shown in Table 13.6.

It is important to identify cash flows not only by how large they will be, but also by their timing, that is, when will they occur. In project economics, it is common to work on an annualized basis, that is, we identify cash flows as a sum for each year. In some cases, it can be relevant to work with shorter periods in project economics. For cash flow analysis as shown in Table 13.6, the presumption is that all payments occur at similar times in all years, which we can take to be either the middle of all years or spread evenly throughout the year.

13.4.2 The time value of money

A key premise in project economics is that the value of money is different depending on when it will be paid. A certain sum of money is considered less valuable if it will be paid in the distant future than if paid sooner. In project economics, the time value of money is considered not primarily in terms of inflation, but for two other reasons: interest and uncertainty.

In the example given above of the machinery, we would invest this year in order to cash in net $20,000 next year and for each following year to the 10th year. But if we expect to get $20,000 in both year 1 and year 10, then the last $20,000 is less valuable to us than the first $20,000. There are two reasons for this:

(1) *Interest*: The first $20,000 can be invested safely in a bank account to earn interest, and therefore it will have grown to a larger amount by year 10 (for instance, at 5% annual interest, $20,000 will grow to $31,026 in 9 years). Alternatively, we can consider the way that the early $20,000 can be used to repay company loans and thus save interest expenses.
(2) *Uncertainty*: What do we really know about the future in 10 years? We may be fairly confident about being able to cash in the $20,000 next year, but the long-term future would normally be considered less certain. Most firms are "risk averse." That means, they would tend to dislike the higher uncertainty associated with the $20,000 in year 10, even if the uncertainty works both ways (i.e., the net cash in year 10 could equally well be higher as lower than $20,000).

The *discount rate* encompasses both interest and uncertainty as compensation for the time value of money. A discount rate is expressed in percent per year, for instance, 10% per year, in the same format as interest rate. Discount rates are however generally higher than the prevailing interest rates, since they also include a compensation for uncertainty. A later subsection provides further considerations on what an appropriate discount factor may be. In project economics, discount rates function so as to reduce the value of future payments. In calculations, they therefore work in the reverse of interest rates.

Table 13.7 shows a calculation where a discount rate of 10% p.a. is used to discount (reduce) the value of cash flows from the machinery investment example. We again have one

Table 13.7: Example of cash flow discounting (based on a 10% discount rate; unit: US dollars).

Year	Net cash flow	Discount factor	Discounted cash flows
0	−100,000	1.000	−100,000
1	20,000	0.909	18,182
2	20,000	0.826	16,529
3	20,000	0.751	15,026
4	20,000	0.683	13,660
5	20,000	0.621	12,418
6	20,000	0.564	11,289
7	20,000	0.513	10,263
8	20,000	0.467	9,330
9	20,000	0.424	8,482
10	20,000	0.386	7,711
11	0	0.350	0
Sum	100,000		22,891 (net present value)

row per year as indexed by the first column. The second column contains the same net cash flows as in Table 13.6. The third column, "Discount factor," contains conversions of the 10% discount rate to annual factors based on the following formula:

$$d_n = (1+i)^{-n}$$

where n is each year, with 0 representing the base year to which values are discounted; d_n the discount factor applicable for year n; and i the discount rate (in this case 10%, expressed as 0.10).

To do this in Microsoft Excel, the formula for the factor for year 1 (placed in cell G30) is:

$$= (1 + \$E\$20)\,^\wedge\,(-A30)$$

where cell E20 contains the discount rate (10% in this case) and cell A30 contains the year, numbered as 0, 1, 2, . . ., with 0 being the base year.

This Excel formula for year 1 yields the value 0.909. It can be copied vertically to the other years, including year 0.

The last column in Table 13.7 shows the discounted cash flows, which are obtained by multiplying each annual cash flow with the annual discount factor from the two preceding columns. For year 1, this indicates that the $20,000 which we expect to obtain in that year has a present value of $18,182. For year 10, the expected $20,000 has a present value of $7711. The question answered here is: what is the value *today* (i.e., in year 0) of the expected $20,000 in year 1 and year 10? The expectation of having $20,000 in year 10 obtains the same value as having $7711 now.

Table 13.7 also shows that a particular sum of money ($20,000) decreases in present value the further into the future it will be paid.

13.4.3 Net present value and internal rate of return

The *net present value* (NPV) of a project is the sum of its discounted cash flows. It expresses the value of all payments resulting from the project, referred to a common point in time (the "present," or at the middle of year 0).

At the bottom of Table 13.7 are two sums: one for the net cash flow and one for the discounted cash flows. The sum of cash flows for all years, $100,000, indicates that the project is expected to generate revenues in excess of expenditures in the amount of $100,000 over its lifetime. An important consideration is however that a large bulk of expenditures has to be made initially, whereas the revenues will come later. This makes the project distinctly less attractive than indicated only by the $100,000 surplus.

The other sum, $22,981 as the sum of discounted cash flows, is a better indication of the attractiveness of this project. It answers the question: how much is the expectation of these future payments, including what we have to invest up front, worth today? This is the pretax NPV of the project. In calculating this, all future expected payments have been converted to their present equivalent amounts, and added up.

One interpretation of the NPV is that the investment will create value for the company in the amount of the NPV. The machine investment project makes our company $22,981 more valuable once the project is embarked upon, bearing in mind that the value of the company also includes the expected future earnings from assets that it has acquired.

The NPV of a series of cash flows depends strongly on the chosen discount rate. If we change it from 10% to 20% in the example, the NPV changes from (positive) $22,981 to (negative) $16,151. A high discount rate implies that little value is assigned to payments in the more distant future. Investment projects generally have lower or negative NPVs if high discount rates are applied, because the future revenues are then not valued sufficiently to make up for the initial outlay.

Decision makers often like to see the profitability of projects expressed in a form that can be compared to interest rates. This would be the *IRR*, which is defined as a discount rate that will yield a NPV of exactly 0.

The NPV of the machine investment project was found to be positive at a discount rate of 10%, and negative at 20%. In between 10% and 20%, there is a particular discount rate that makes the NPV of this project exactly 0. This is the internal rate of return (IRR). In this case, it works out to 15.1%. It indicates an expectation that the project will generate similar returns as putting money in the bank and obtaining 15.1% effective interest.

Besides NPV and IRR, the following three measures of economic performance are often used in project assessment. They would normally be used as supplementary information, more seldom as stand-alone criteria for investment decisions.

- *Profitability index*: A ratio obtained by dividing the NPV by the original investment.
- *Payback*: The time taken from start of investment to recovering the original investment as cumulative cash flow.
- *Maximum exposure*: The largest cumulative negative cash flow early in the project (often, but not always, the same as the original investment amount).

13.4.4 Dealing with inflation in project economics

The 1970s and 1980s were a time of high inflation for many of the world's economies. Inflation has since fallen back, and in the early 2000s it was usually less than 4% in the major economies. This has tended to simplify things in project economics. However, as long as there is some inflation, one needs to be clear about how we account for it in project economics, especially since these calculations often extend many years into the future.

A distinction is made between *nominal monetary terms* and *constant monetary terms* as modes of project economic calculations. A calculation in nominal monetary terms means that all monetary values in the calculation are to be taken literally, regardless of when they arise. A calculation in constant (sometimes referred to as "real") monetary terms displays monetary values by constant purchasing power. The constant purchasing power must be referred to a year, which should be the same year as used for discounting purposes. In a calculation in constant monetary terms, if we assume 3% annual inflation and show a value of $100 in year 2, then this represents $106.09 (=$100 × 1.03 × 1.03) in actually expected money (nominal money).

There is a need for consistency in applying either nominal or constant monetary terms. In a constant-terms calculation, the discount rate must also be understood in constant monetary terms, that is, a percentage over and above the assumed inflation. There is a potential complication in calculating taxes, because depreciation, which has an impact on taxes, usually must be calculated in nominal terms. The assumed inflation should be linked to the currency in which the calculation is done, that is, in a calculation made in US dollars, it is appropriate to use inflation as assumed in the United States (usually the consumer price inflation), even if the project takes place in a different part of the world where local inflation may be very different.

The cash flow calculation in Table 13.6 has the appearance of being in constant monetary terms, since revenues and costs do not increase in time. In a low-inflation environment such as the early 2000s, these could be realistic cash flows even in a nominal calculation. Many traded goods saw little or no price increases, due to competitive pressures and productivity gains. Futures prices for crude oil are often *backwardated*, that is, oil for delivery in the far future is traded at lower prices than oil for earlier delivery, which is the opposite of what one would expect under inflation. It can also be reasonable to assume no increase in labor costs under the assumption that salary increases will be offset by productivity gains.

Sometimes, project economics involve two or more different currencies, and assumptions about exchange rates must be incorporated in the economics calculation. The inflation assumed for each currency can be quite different. At least in the long term, one would expect exchange rates to adjust broadly to differences in inflation, so that the currency that exhibits high inflation will normally be on a falling trend against the low-inflation currency. In the short and medium terms, this is not necessarily the case.

13.4.5 Choice of discount rates and criteria for investment decisions

Project economics are usually done for the purpose of preparing for investment decisions. Many firms use the NPV calculation as the main economic criterion for deciding on investments. A discount rate is then applied which reflects the firm's required return on such investments. An investment is considered profitable if this discount rate results in a positive NPV. The profit-optimizing firm should select a portfolio of investments which maximizes positive NPV overall (using NPVs as the criterion for selecting investments sometimes results in different decisions than using IRRs; the NPV approach is generally recommended in theoretic literature).

The choice of discount rate(s) to be used in project economics, and, more generally, what financial criteria to apply for investment decisions, is often a matter of corporate policy. Large organizations usually issue corporate instructions in this regard.

Financial theory provides guidance in how firms can determine the required returns to be used as selection criterion for investments. It starts from the premise that firms obtain capital from financial markets where investors demand returns based on what they can obtain from other projects with similar uncertainty. Investors demand compensation for accepting uncertainty. A distinction is made between so-called "systematic" uncertainty and individual project uncertainty. Investors can largely offset individual project uncertainty by spreading their funds on many different assets with uncorrelated uncertainties, and therefore should not need to demand any significant premium for that type of uncertainty. For a full explanation of this issue, refer to a good textbook on financial management and capital budgeting (e.g., Brigham and Ehrhardt, 2007).

It is sometimes argued that public entities – state agencies, public utility firms, etc. – can and should apply lower discount rates than usually done by private-sector investors, including firms operating internationally. There is some merit in this on the basis that the comparison to anticipated returns in equity markets is less relevant for such entities, which pursue other goals than profit maximization in the purely financial sense. Also, public entities may well perceive uncertainty differently from foreign investors when investing in long-term assets within their own country. In some countries, the required return rate for public investment in infrastructure and utility services has been the subject of much politicized debate, resulting in public entities applying explicitly lower discount rates than normally would be expected from private investors.

13.4.6 Assumptions and uncertainty

The outcome of a project economics analysis depends critically on the assumptions entered in the calculation. Section 13.3 introduced cost estimation, including comments on handling uncertainty in that context. Other parameters required for the economic analysis of petroleum projects are also uncertain, such as production profiles and, not least, oil prices.

For important macro-economic parameters such as oil prices and currency exchange rates in the future, firms usually issue corporate guidelines which affiliates are required to follow.

Unfortunately, professionals from different disciplines tend to assign different meanings to the term *risk*. Technical professions tend to associate "risk" with the possibility of an outcome that is worse than a norm (including the risk of accidents as reviewed in Chapter 10). Economists tend to associate "risk" with the possibility of an outcome that is different from a norm, and not necessarily worse. In this text, the former meaning is used (i.e., risk is the danger of an adverse outcome), while *uncertainty* is used for any possibility of different outcomes.

The concept of NPV takes account of uncertainty, which is incorporated as an increment (economists refer to this increment as a "risk premium"; it would be more consistent with the terminology used in this text to call it an "uncertainty premium") in the discount rate above ordinary bank interest rates. However, this is a valid approach only if the uncertainty is balanced relative to the assumptions used. Acknowledging the uncertainty of future cash flows, the NPV of a project should be based on cash flow that represents the mean of a probability distribution function, that is, the expected value as indicated in Figure 13.3.

When hearing statements about the riskiness of a project, one should be alert to potential misunderstandings. An economist may interpret "high risk" to mean that the project has a wide range of possible outcomes, but that the calculated NPV represents the mean probability of possible outcomes, so that the chances of better or worse outcomes are balanced. A practically minded manager may interpret "high risk" to mean that the calculated NPV represents a target based on a plan, but that there is a rather large chance of cost overruns, production failures, or other negative deviations, while the chance of a better outcome is remote.

Conceptually, the discount rate is meant to reflect the first kind of uncertainty (large but balanced deviation around a mean probability), but not the second kind (overwhelming probability of more unfavorable outcome, i.e., a "risk" in the sense used here). For a calculated NPV to be valid as a criterion for decision making, it must be based on the first kind of assumptions, that is, representing means of probability distributions, so as to provide a balanced uncertainty. Intuitively people may have some problems with this, because the mean of probabilities for assumptions is not necessarily what is described in project plans. It may not even reflect any credible story about what is supposed to take place, but rather the abstraction of a statistical mean of possible events.

There are ways in which uncertainty associated with a project can be analyzed more explicitly. The two most relevant methods are *sensitivity analysis* and *Monte Carlo simulations.*

Sensitivity analysis is a calculation of project economics (or some other parameter) under different assumptions. The solution reflecting mean probabilities (or a project plan) is referred to as the *base case*. A simple way of making sensitivity analysis is to vary one input parameter at a time, and calculate what it does to the end result. One may, for instance, vary the assumed oil prices, Capex, reserves, etc., with ±25%, and calculate the effects on the project NPV and IRR.

A more sophisticated kind of sensitivity analysis is to assume a different outcome of a specific event, and calculate what it will do to project economics. For instance, if a particular lifting barge that would suit the project will not be available at the required time, then alternative plans will be set in motion affecting the project in several ways. This might be called a *Plan B analysis.*

Where sensitivity analysis is made, it may be justifiable to relax the requirement of base case NPV reflecting mean probabilities, and instead use assumptions reflecting specific events or plans that can be described. It is however necessary to communicate clearly what calculated NPVs represent, and to understand the uncertainties of a project as a basis for making decisions.

Monte Carlo simulation is a procedure for generating a probability distribution of an output parameter, when that output parameter is dependent on several input parameters for which probability distributions have been defined. The procedure, usually done with specialized computer software, calculates the output parameter a large number of times, each time assigning randomized values to the input parameters based on their probability distributions.

13.4.7 Complex cash flows: beware of the dual IRRs

Consider the following case: an offshore oilfield has been operational for several years, and is still expected to produce 100 million remaining barrels over the next 15 years. The production rate is presently constrained by processing and water injection capacities. A proposal is made to spend $120 million to upgrade capacities, resulting from which production would increase by 20% right away. It would however not change the total amount of oil ultimately produced from the field, but it would reduce the time taken to deplete the field from 15 to 12 remaining years.

Table 13.8 shows a summary of cash flows that may result from this. The cash flow of the process upgrading project, shown in the last column, is derived as the difference between two

Table 13.8: Incremental cash flow for upgrading oilfield processing (example; unit: million dollars).

Year	Net cash flow without process upgrading	Net cash flow with process upgrading	Project cash flow: effect of upgrading
0		−120	−120
1	280	335	55
2	280	335	55
3	280	335	55
4	280	335	55
5	280	335	55
6	280	335	55
7	250	275	25
8	220	185	−35
9	175	125	−50
10	130	65	−65
11	100	35	−65
12	70	5	−65
13	40	0	−40
14	25	0	−25
15	10	0	−10
16	0	0	0
Sum	2,700	2,580	−120
NPV at 15%	1,353	1,372	18

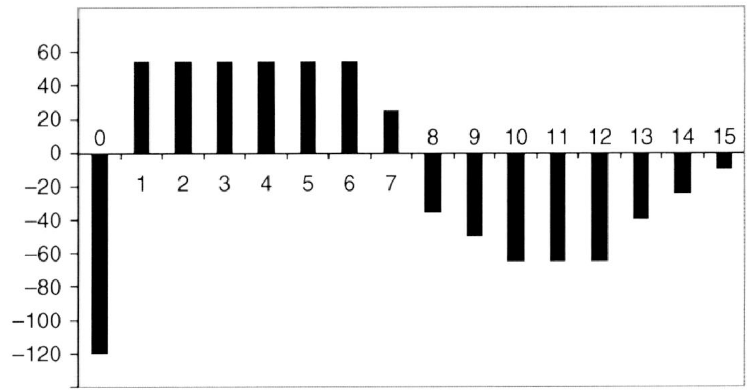

Figure 13.7: Project cash flow from Table 13.7.

cash flows: the total cash flow from running the field as it is for the remaining 15 years and the total cash flow from the field if the investment is made. Figure 13.7 shows the project cash flow as a histogram.

Running this in MS Excel obtains an IRR of 9% from the last column. This strikes as odd since a positive NPV of $18 million is shown when discounting at 15%: in normal investments, we get negative NPVs when discounting at higher rates than the IRR, and positive NPVs at lower discount rates. Here, the reverse is the case.

The anomaly is due to the pattern of cash flows resulting from this project: cash flow changes direction from negative to positive, and then back to negative. After the initial outlay (−120), there are several years of positive cash from the increased oil production. But ultimate oil recovery was assumed to be unaffected, so there will be some years toward the end where oil production will be less as a result of the project than it would have been without it, hence the negative cash flows in later years.

Contrary to ordinary investment projects, this project does not generate extra revenue. Its possible merit is in accelerating given total revenue, so that it is obtained earlier than would be the case without the project.

It is possible to calculate two IRRs from this cash flow series: one at 9% and one at 34%. Neither is very helpful in determining the merits of the project. In particular, beware of any IRR (the 9% in this case) that has the property that the NPV at a discount rate slightly lower than the IRR is negative. Such an IRR will increase if the project deteriorates, and does not indicate project returns in the ordinary sense.

Also in a project like this, it makes little sense to set the discount rate high in order to compensate for uncertainty. The NPV in this case turns out higher at 20% discounting than at 15% and 10% (but lower again at 25%).

Ordinary NPV calculation gives an indication of whether this project has merit or not, but one should be careful about not using a too high discount rate, which would put the project in a too favorable light. In addition to running NPV at the firm's ordinary discount rate, it is suggested to also run NPV at a discount rate corresponding to the firm's cost of borrowing, which is lower than the ordinary discount rate (make sure there is consistency in the currency applied). If this yields a negative NPV (which it may well do, even if the other NPV is positive), then the project does not have clear merit unless it is considered to reduce the firm's overall exposure to financial risk.

13.4.8 Criticism and new approaches to project valuation

The reversed cash flow case in the previous section points to some general problems with the discounted cash flow approach to project valuation. Laughton (1998) highlighted some perceived shortcomings of the traditional approach to project economics with special relevance for petroleum. Conventional cash flow discounting as a basis for investment decisions was believed to frequently result in the following errors:

(1) underestimating the relative value of long-term reserves;
(2) bias toward building too much capacity, leading to inefficiencies when production declines;
(3) undervaluing future costs, leading to a preference for low-quality equipment with low initial costs but higher future maintenance costs;
(4) inability to evaluate the unique risk profiles represented by similar projects under different fiscal regimes;
(5) not valuing flexibility.

The first three points can be related to the level of discount rates (required returns) selected. Using too high discount rates for NPV-based investment decisions can certainly lead to those problems. Building an uncertainty factor into the discount rate implies that uncertainty increases exponentially with time. This may be a fair representation of certain types of uncertainty, but not all.

One simplified remedy for this problem would be to apply a lower discount rate than conventionally used, and then applying a requirement that the NPV thus calculated for a project must be positive above a certain ratio to the initial capital commitment (the profitability index; see Section 13.4.3). This would tend to favor investments with long-term benefits over those with short-term benefits, compared to the conventional approach. Laughton (1998) proposes a more elaborate approach which involves decomposing the cash flow into individual components to which different discount rates may be applied.

One example can be related to the oilfield referred to in Figure 13.7. In the base case (no investment), the field is assumed to keep producing for another 15 years, and certain assumptions were made about the oil price and the production rates. These are two uncertain variables, for which probability distributions can be generated. If one or both of them turn out less favorable than projected in the latter years, it will hurt the firm's cash flow during those years. But the firm can surely have the option to close the field down earlier than after the 15th year, should such circumstances arise, which would tend to soften the impact of the reduced revenues by eliminating operating costs for some period. The possibility of closing down at a different time than originally foreseen is a *real option* for the oil firm, and has value to it as a way of responding to uncertainty.

A real option is a recognized ability of a firm to enhance value through flexibility, that is, the ability to act in different ways in the future so as to adjust to outcomes that are initially uncertain. Real options have been a prominent feature of economic and management literature especially in the 1990s (Luehrman, 1998). There are techniques developed for assigning values to real options, and to describe ways in which they can be used in response to unfolding events. The most common such technique is the *decision tree*, which features nodes of uncertainty and decisions.

The implication of the real-options approach is that firms should sometimes be willing to pay an extra cost in order to acquire flexibility; as such, flexibility can become quite valuable in the future. Astute business people have of course always made such decisions based on more intuitive perceptions, but the real-options literature offers some more formalized approaches.

13.4.9 Economics and strategy

Net present values and the related concepts are a guide to making decisions on investments, with the implication that investments with a positive NPV add value to the firm. Those who have an interest in the discipline of business strategy, will argue that such quantitative measures of profitability cannot be the only criterion for how the organization should deploy its resources: investments must above all be consistent with the organization's strategy.

Competitive strategy is concerned with how firms can develop the ability to survive and prosper in competitive environments. The development of strategy for a firm is usually a deliberate process led by its senior management and Board, and resulting in formal decisions. In the strategic perspective, decisions are viewed as a pattern which must be coherent. Individual decisions are then evaluated not only by their NPVs, but above all by their contribution to moving the firm in the intended strategic direction. The discipline of strategy is mainly qualitative rather than quantitative in orientation, dealing in terms such as core competence, strategic intent, entry barrier, and sustainable competitive advantage – terms for which there are no formulas in Microsoft Excel.

Decision making in organizations has a strategic dimension in addition to the quantitative dimension dealt with in the preceding chapters. Sometimes investment opportunities should be foregone in spite of showing attractive economics, because they would detract from the firm's strategic direction. Sometimes investments should be made in spite of showing unattractive economics when viewed in isolation, but contributing essentially to building strategic resources. And sometimes, projects are misguidedly labeled "strategic" as a poor excuse for pursuing someone's unprofitable pet idea.

Strategies are however not successful if they fail to meet the organization's financial objectives. Applied in that context, NPV and related techniques are relevant criteria also for strategic decision making.

13.5 Probability distribution

Planners of petroleum developments often need to have an opinion about conditions that are important for the development, but which cannot be known for certain. The uncertainty may have different causes: some parameters will be determined only in the future, such as the market prices for oil which will provide the sales revenues for a project. Other parameters are determined but not fully observable to humans, and therefore have to be treated as uncertain, such as underground reservoir conditions.

Probability distribution is a structured approach to dealing with uncertain parameters for the purpose of numerical analysis. As a field of science, it is related to statistics.

13.5.1 Introducing probability and normal distribution

In business analysis, we often have to make assumptions about something that we do not know for certain. Consider the following questions:

(1) By how many dollars per barrel will the price of Brent oil rise or fall over the next 30 days?
(2) What will be the price of light, sweet crude oil in 2050, on average?

(3) Will the United Kingdom be a net importer of oil in 2020?
(4) How much will it cost to drill a well?
(5) What is the total recoverable amount of oil in the Brent oilfield?

Any of the above questions might be relevant for a petroleum business. The answers cannot be known for certain, at least not yet. In business analysis, it is quite common (but not necessarily recommended) to take a *deterministic* approach to such questions, that is, to figure out one reasonable answer to each question and base the analysis on that number only, ignoring the uncertainty. The alternative is to treat these parameters as *stochastic* variables, that is, uncertainties.

The terms *deterministic* and *stochastic* are basically applied to processes. A stochastic process involves uncertainty, whereas a deterministic process does not. Stochastic analysis involves variables that are treated as uncertain. Instead of being given just one assumed value, such variables are given a *probability distribution*, that is, a mathematical description of the likelihood of different values that the variable can have.

Parameters such as those reflected in the five questions above might be required inputs for a decision-making process. Sophisticated software tools for economic analysis usually have a facility for treating such parameters stochastically. The uncertainty can then be incorporated into the analysis, not just assuming one answer for each parameter.

Turning to the first question raised above: an opinion is needed on how much the price of Brent crude oil will change over the next 30 days. The basic view is that the price will fall by \$2. Section 13.5.3 provides some ideas on how such a view may be generated. It is also recognized that this is far from certain. The price may fall further, or it may instead rise.

Figure 13.8 gives an example of how such an assessment of probable price changes might be described. It is a bell-shaped curve which peaks at −2, which is the "best estimate" of what will happen. Further out to the left, the curve shows lower values, indicating that such larger price drops are possible but distinctly less possible. To the right, there are also lower values indicating the lesser likelihood of price increases. We see that the curve on both sides never quite reaches zero. This indicates that there is no definite limit to how much the price of oil can rise or fall, but changes beyond the shown range have very small probabilities.

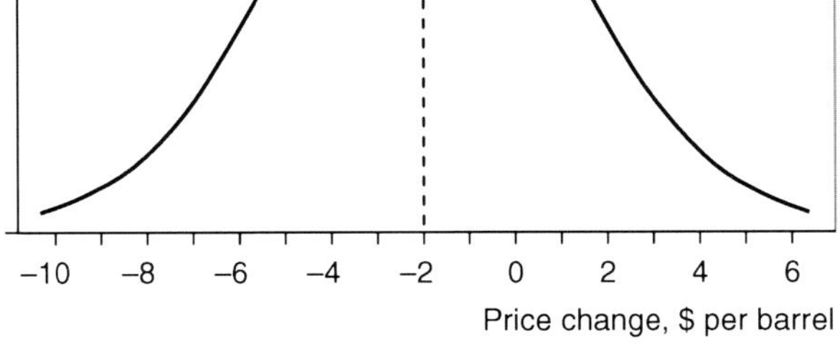

Figure 13.8: Probable price change of oil over next 30 days (normal distribution displayed as a density function).

A bell-shaped curve like Figure 13.8 is called a *normal distribution*. It is an important concept in statistics, and often used as basis for probability analysis. Many phenomena in life which are quantifiable and can vary tend to be variable in ways that resemble a normal distribution.

To arrive at a normal distribution like the one in Figure 13.8, two parameters are required: first, the *mean*, that is, the view of what is most likely to happen, which in this case is a price change of −2$ per barrel; second, the *standard deviation*, which is a measure of how variable and uncertain this phenomenon is. In this case, a standard deviation of $3 per barrel was used.

The *mean* is the average value of a population, a sample, or a probability distribution, when all values are counted equally. Expressed mathematically:

$$\bar{x} = \frac{\sum x}{n}$$

where x is each observed or expected value, $\bar{x}$ the mean of all x, and n the number of x.

The *standard deviation* is a measure of the tendency for the individual values to differ from the mean. A large standard deviation implies a large spread between values or a large uncertainty. Its formula (which not everyone needs to memorize) is:

$$\sqrt{\frac{\sum (x-\bar{x})^2}{n}}$$

The standard deviation has the property that in a normal distribution, about 68% of all occurrences fall within the range of the standard deviation on both sides of the mean (the following applies to normal distributions generally: approximately 68% of all occurrences are within ±1 standard deviation from the mean, 95% are within ±2 standard deviations, and 99.5% are within ±3 standard deviations). In this case, that range is from −5$ to +1$ per barrel, as possible changes from today's oil price (we chose −2 as the mean and 3 as the standard deviation. This gives a range from −5 to +1, because −2 − 3 = −5 and −2 + 3 = 1). This implies, there is a likelihood of about 0.68 (i.e., 68%) that the oil price in 30 days will be within −5$ to +1$ per barrel as a change from today's price. There is therefore a probability of 0.32 that the price of oil will end up outside that range (the combined probability of all possibilities is by definition 1 (or 100%); the number 0.32 referred in the text is therefore arrived at as 1 − 0.68 = 0.32, where 0.68 was the probability that the price will end up within the described range). The normal distribution is symmetric around the mean. The probability of 0.32 therefore has two equal parts: a probability of 0.16 on the negative side (price falling more than $5) and also a probability of 0.16 on the positive side (price rising more than $1). In summary, we are implying the following probabilities about the price of oil 30 days from now:

- probability of 0.16 that the price will fall by at least $5 per barrel;
- probability of 0.16 that the price will rise by at least $1 per barrel;
- probability of 0.68 that the price will be within the range of −5 to +1$ from today's price.

13.5.2 Cumulative probability distribution

Many people find that a curve like Figure 13.8 illustrates two points well: a price change of −2$ is considered likely, and price changes both lower and higher than −2 are possible, but less likely the further away from the mean expectation of −2$. But this graph is not easy to work with beyond that. It has no vertical scale, for which additional assumptions would be necessary.

The bell-shaped curve in Figure 13.8 is also called a *density function*. Figure 13.9 contains the same information as Figure 13.8, but shown as an *inverse cumulative distribution graph*. The vertical axis is now a scale from 0 to 1, which indicates probabilities. A probability of 0 for an outcome means that it certainly will not occur, and a probability of 1 for an outcome means that it will certainly occur. The graph curve indicates, for each price change value found on the horizontal scale, the probability that the price in 30 days will be at least that much.

The vertical dotted line at the price change value of −2, which is the chosen mean, crosses the graph curve at the cumulative probability value 0.5. The implication is that there is a probability of 0.5 (50%) that the price of oil in 30 days will be not lower than −2$ per barrel as a change from today's price.

The graph in Figure 13.9 shows an *inverse* cumulative distribution, because it starts with something that has a very high probability (that the price in 30 days will be not lower than $10 below the current price). The cumulative probability curve falls as the horizontal scale shows higher figures for the oil price, because this lowers the probability of the price staying above those levels.

Here is an example of how the graph in Figure 13.9 may be read. One might ask the question, what is the probability of the oil price rising, even though the best estimate is that it will fall by $2 per barrel? Identify the value 0 on the horizontal scale. This represents today's price of oil (zero change). Use a ruler to make a vertical line up from the value 0. It hits the graph curve near the horizontal line indicating 0.25 as a cumulative probability. This implies a 0.25 chance that the oil price in 30 days will be higher than today's price. Correspondingly, there is a 0.75 chance that it will fall.

This probability of 0.25 for a price rise results from the two parameters which were assumed: the mean expectation of a $2 fall in prices and the standard deviation of $3, as a measure how uncertain we consider the estimate to be. A bigger standard deviation would make the curves in Figures 13.8 and 13.9 flatter and more extended, and would increase the probability of a price rise even with the same mean expectation of a $2 drop.

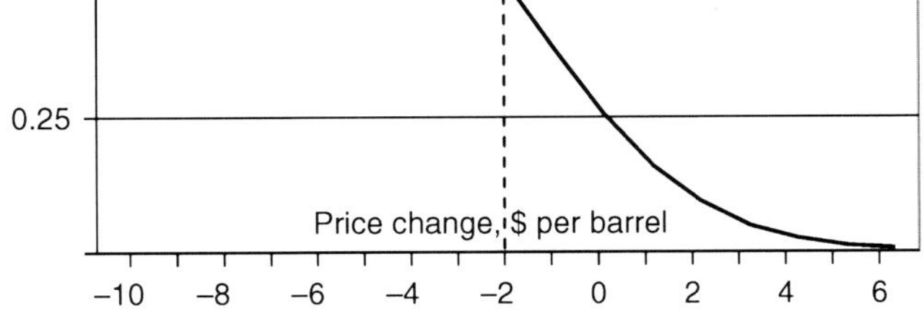

Figure 13.9: Probable price change of oil over next 30 days (normal distribution displayed as an inverse cumulative distribution).

13.5.3 Generating a view on expected value and uncertainty

Generating a view of the mean and the standard deviation in a case like the example above is essentially a judgmental process. It should be supported by professional knowledge about the phenomenon in question. This knowledge may come from empirical (actual) observations of the past, or from theoretic models. Here is how a 30-day oil price forecast might be generated, in three steps.

As a first step, define precisely the variable to analyze. In this case, clarification is needed on which oil price is referred to. For Brent crude oil, there are essentially two types of readily observable prices: there are the prices from future trading, which for Brent crude oil traditionally have been on the International Petroleum Exchange (IPE) in London (www.theice.com). Alternatively, select the prices that are formed in direct trade between buyers and sellers of oil cargoes, as reported daily by agencies such as Platts (www.platts.com) or Petroleum Argus (www.argusonline.com). In this example, the selected parameter will be the Dated Brent quotation by Platts. Since this quotation comes as a range (for instance, $105.20–105.25, depending on what happened in the market during that particular day), one may, for instance, select the high end of the range (or the low end or the average, as long it is clarified). The Dated Brent quotation reflects prices that are agreed today for deliveries of oil to be made within a couple of weeks. By selecting this quotation, we have also specified the quality of oil which will underlie our analysis (the Brent Blend, whose quality is well known to market participants) and the mode of delivery (FOB at certain loading terminal in the North Sea), both of which affect the price of oil.

The second step is to form a mean expectation about the outcome of the defined variable, in this case Platts' Dated Brent quotation 30 days hence. By following the market closely, one may form an independent view based on short-term supply balances, evolving political tensions, the weather in important consuming areas, and other factors which may influence the outcome. This process may be supported by market reports and the views of credible analysts. For oil prices, one may also look at forward quotations. Platts and Argus quote not only prices for prompt delivery (Dated cargoes), but also prices agreed now for delivery in the following few months, which are usually different from the dated prices and reflecting expectations in the market.

The third step, and the trickiest, is to form a view of the uncertainty. For now, assume that the uncertainty can be expressed in the form of a normal distribution, but that will not always be the case. In later subsections, other distributions will be reviewed.

Uncertainty is expressed as the standard deviation, in the form of a number. The standard deviation does not lend itself easily to judgmental or intuitive estimation, so an assessment should be backed up by relevant facts. Records of past events of a similar nature and their variability can be a useful reference. This would certainly be possible for the price of Brent crude oil, for which time series of all past quotations can be obtained.

A *population* consists of all elements of the relevant set. If the issue of interest is exploration wells in the North Sea, then the population is all such wells ever drilled in the North Sea.

A *sample* is a selection of the population, such as a smaller number of wells that have been drilled in the North Sea. For practical reasons, a sample is often studied in order to make conclusions about the entire population, provided that a sample has been selected that represents the population well. Preferably it should be obtained by *random sampling*, that is, a process by which all members of the population had the same chance of becoming members of the sample.

It is common to use different mathematical symbols representing the mean and the standard deviation, depending on whether they apply to a population or a sample:

μ (Greek small "mu"): mean of a population;
σ (Greek small "sigma"): standard deviation of a population;
$\bar{x}$: mean of a sample, when observed values are denoted as x;
s: standard deviation of a sample.

The spreadsheet function AVERAGE(range) can be used for calculating the mean, where range is the series of cells containing all your monthly changes. For the standard deviation, use the spreadsheet function STDEVP(range). The *P* in the function name is appropriate when working with the entire population of past cases. In other cases, the analysis is made on a sample rather than the full population, in which case the standard deviation should be calculated with the formula STDEV(range).

13.5.4 Continuous and discrete distributions

The discussions above concerned variables that had continuous probability distributions. That is, the variable (such as the price of Brent crude in 30 days, or the total recoverable quantity of oil in a field) can take an infinite number of different possible values, in infinitely small steps along the scale.

The alternative to continuous distribution is discrete distribution. Here, the possible outcomes are restricted to certain specified values. An example of a discrete distribution is the result of rolling a die. The outcome can only be 1, 2, 3, 4, 5, or 6 – never anything between. The probability of obtaining each of these specific values should be exactly 1/6.

Another example of a discrete distribution is the question (3) in the beginning of this section: will the United Kingdom be a net importer of oil in 2020? As posed, this question has only two possible outcomes, Yes and No (which we can express numerically as 1 and 0). An informed analyst would probably not assign a 0.5 probability to the Yes outcome, but assign a higher or lower probability based on analysis of the oil production potential and domestic consumption projections for oil products.

The question can also be posed differently: what will be the net imports of oil to the United Kingdom in 2020? This question can be answered with a continuous probability distribution.

In continuous distributions, probabilities are not assigned to specific values, only to ranges of values. If the distribution is truly continuous, then the possibility of the oil price falling by *exactly* \$2 per barrel has no probability. If pressed for an answer, a statistician might tell you that it has a probability of zero. That may come across as frustrating since this was supposed to be the most likely outcome. The reason for the frustration is then that the outcome of −2\$ is just one of infinitely many outcomes, another one of which is −2.000000000000001\$, which is not \$2. So the chance of this hitting exactly −2\$ is infinitely small (which may be treated as zero), even if this is where the probability distribution is at its densest.

In a continuous distribution, therefore, no probability is assigned to the exact value of −2\$ per barrel. But a probability can be assigned to a range of values represented by the index value −2\$. That range may be, for instance, from −1.50 to −2.50.

There is a clear and important distinction in theory between continuous and discrete distributions, but less so in practice when dealing with available figures. Theoretically, when people agree to buy and sell oil, they might agree any price that can be expressed as a number on an invoice. In practice, outright prices agreed and reported for oil are usually in full dollar–cents per barrel. When the price of Brent crude in 30 days is reported by Platts, it will be in the form of a figure expressed in full cents, because that is how Platts reports. This makes it part of a discrete distribution of possible prices, not continuous. However, a discrete distribution like this, with a dense set of index values ($105.00, $105.01, $105.02,...), can for most purposes be treated like a continuous distribution.

13.5.5 Normal vs. lognormal distribution

Turning to the second question posed in Section 13.5.1: what will be the price of light, sweet crude oil in 2050, on average? This is much further into the future than conventional market analysis will cover – never mind that conventional market analysis often gets it very wrong also when it comes to predicting short-term oil prices. There are some publications that try to describe very long-term developments in global energy (for instance, Shell's long-term scenarios, at www.shell.com), but they will not give any solid and precise price forecasts. There are obviously enormous uncertainties with regard to new technologies for making energy available, and other things that will affect the outcome.

One might come up with the view of $50 per barrel as a best estimate, but with such a large uncertainty that both $5 and $500 per barrel are also considered possible outcomes – and equally so. The probability distribution might look like Figure 13.10. This looks much like the normal distribution in Figure 13.8, but the horizontal scale is of a different nature. The index values $5 and $500 are now placed in equal distance from the middle value of $50. Arithmetically this does not hold, since the difference ($500 − $50) is not the same as ($50 − $5). But the distances are the same in the proportional sense: $500/$50 = $50/$5. This makes it a *logarithmic scale*. The distribution that is normal (with the symmetric, bell-shaped form) on a logarithmic scale is called a *lognormal distribution*.

Lognormal and similar distributions are often relevant in stochastic modeling for economic purposes. This is because many phenomena that we need to analyze have a lower boundary but no upper boundary. This is certainly the case for such important parameters as the future price of oil, the amount of oil in a reservoir, and the number of drilling hours needed to reach the reservoir. None of these values will ever be negative, so zero is a lower boundary (and even that is extremely remote). At the high end, there is usually no absolute limit, only diminishing probabilities.

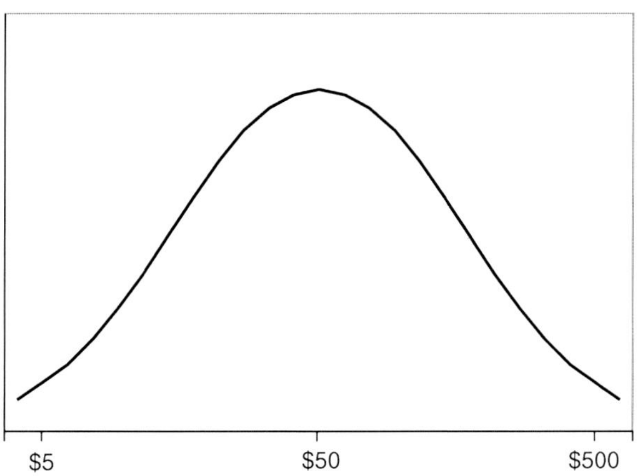

Figure 13.10: Expectations for the price of crude oil in 2020 (displayed as a lognormal density function on a proportional scale).

The symmetry which we assumed for the regular, normal distribution is strictly speaking not there. The normal distribution assumes no lower or higher limit to possible values – only that the values far from the mean will have tiny probabilities. In reality, probability distributions for variables such as the future price of oil should have an absolute lower limit, because the price will surely never be less than 0. In the previous case on the 30-day price forecast, this effect could be ignored without much error, because nearly all of the probability distribution was far away from the lower boundary of 0 (assuming oil prices as of 2008).

When considering the price of oil in 2050, this issue cannot be ignored, because $500 oil is a significant possibility, whereas oil price of −400$ is not a relevant possibility. At first sight, the lognormal distribution in Figure 13.10 also appears to have no lower boundary – the curve will never reach zero no matter how far to the left or right we extend it. But the horizontal axis values also never reach zero if extended to the left; they just become ever smaller positive numbers. Therefore, the value 0 is indeed a lower boundary for this lognormal distribution.

When shown on a linear scale, the lognormal distribution is strongly skewed to the left (Figure 13.11). This graphical presentation also reveals that the three common measures of central tendency, the mode, median, and mean, are all different in this distribution (at around 20, 50, and 80, respectively). In the normal distribution, these three values coincide.

Following are the three measures of central tendency used with probability distribution functions:

- The *mean* is the average of the population, that is, average of all values in a discrete distribution.
- The *median* is the middle value, that is, the population is equally large above the median and below the median. It is also called the 50 percentile.
- The *mode* is the value that has the densest distribution, that is, the peak of the distribution curve.

Lognormal distributions can be handled numerically by converting all values to logarithms, and then associating a mean and a standard deviation with the obtained values. Microsoft Excel has a function for cumulative lognormal distribution:

LOGNORMDIST(*n*; mean.ln; sd.ln)

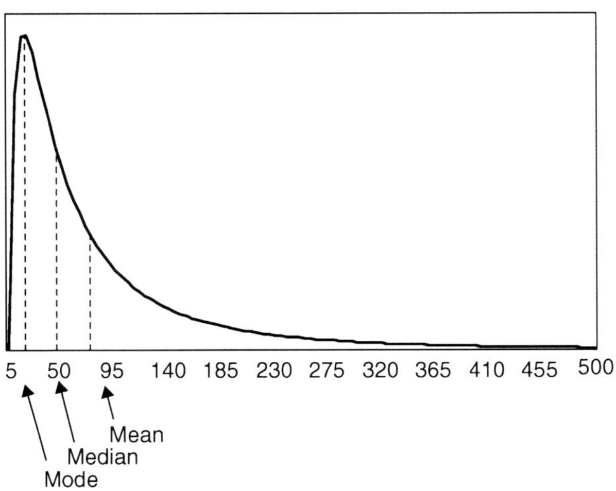

Figure 13.11: Expectations for the price of crude oil in 2020 (displayed as a lognormal density function on a linear scale).

where n is the index value for which the cumulative probability is calculated, mean.ln the mean of $\ln(x)$, sd.ln the standard deviation of $\ln(x)$, and $\ln(x)$ the natural logarithm function applied to the population values.

Figures 13.10 and 13.11 were generated using the following values: mean of $\ln(x) = \ln(50)$; standard deviation of $\ln(x) = \ln(3)$.

13.5.6 Triangular distribution

Sometimes things need not be so complicated. Imagine the following conversation between a planning analyst and a drilling superintendent working on a hydrocarbon prospect somewhere offshore Africa:

Drilling superintendent: We've got plans ready for drilling that appraisal well.
Planning analyst: How much do you expect it will cost?
Drilling superintendent: 9.4 million US dollars.
Planning analyst: Are you sure?
Drilling superintendent: No, but that is our budget estimate. Things can go against us, and then it will cost more.
Planning analyst: How much could it cost in the worst case?
Drilling superintendent: I'd say, certainly not more than 14 million. That's really the worst case.
Planning analyst: And the best case?
Drilling superintendent: 7 million. That's drilling at more than record speed with no hiccups whatsoever.
Planning analyst: Great – can you give me the standard deviation of the probability distribution of the drilling costs expressed as natural logarithms and based on a lognormal distribution?
Drilling superintendent: No.

The analyst might have dropped her last question, up to which the two were getting on well. If the superintendent is an experienced professional with a good judgment of such things, he had already given the analyst a suitable basis for a probability distribution, as shown in Figure 13.12. This is a triangular distribution defined by the three numbers stated by the drilling superintendent: minimum (\$7 million), most likely (\$9.8 million), and maximum (\$14 million).

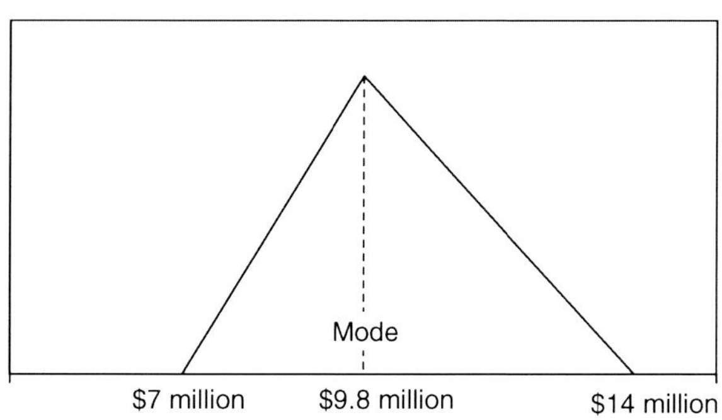

Figure 13.12: The expected cost of drilling a well (displayed as a triangular density function).

This methodology is simple and often appropriate, bearing in mind that real-life probability distributions are often based largely on subjective judgment. One should however be aware of certain statistical properties of this kind of probability distribution.

Even if there would be some chance that the cost may fall outside the assessed minimum–maximum range (as numerous drillers have found out the hard way), the suggestion here is to disregard that possibility. As shown in Figure 13.12, the superintendent's statement about $9.8 million, his budget estimate, was taken to be the *mode*, that is, the index value with the highest probability density. Note that this is not the same as the *mean* or the *median*, both of which are higher than the mode in this case because of the way the distribution is skewed. By this distribution, the well cost will probably end up higher than the superintendent's estimate. This is not in contradiction to accepting his best estimate of what the cost will be.

This implies some interpretation about what the superintendent meant by what he said, and what he might have a basis for saying. To some extent, it depends on the culture and norms of the organization with respect to cost control and operational responsibility. As a practically oriented line manager, the superintendent probably has his mind set on $9.8 million as a reasonable number that he should be able to stay within, based on the information he has received about the geological conditions. But he also recognizes the possibility of the costs ending up considerably higher, and to a lesser extent lower, than his estimate, for reasons of which he has only partial control.

13.5.7 Other probability distributions

The simplest probability distribution, or rather, lack of such, is the *constant*. Specifying a variable as a constant means just assigning one value to it, not allowing for any possibility that it may turn out differently.

A *uniform distribution* is also rather simple: it is defined by a minimum and maximum, and the probability density is the same across that entire range (Figure 13.13). This may be appropriate if someone gives a cost estimate of "anywhere between $7 million and $14 million" – without any clue as to what may be more or less likely within that range.

Multiple percentiles distributions can be used if there is a direct basis for knowing the possible outcomes associated with probabilities. One example is the *7-percentile distribution*, which allows for specifying cumulative probabilities at the percentiles of 0%, 5%, 25%, 50%,

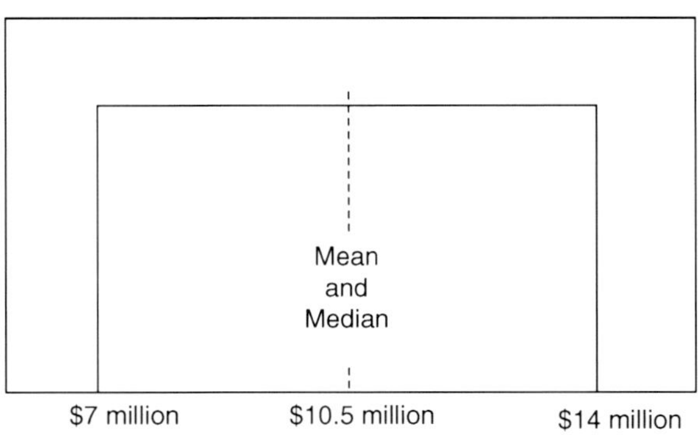

Figure 13.13: A uniform probability distribution.

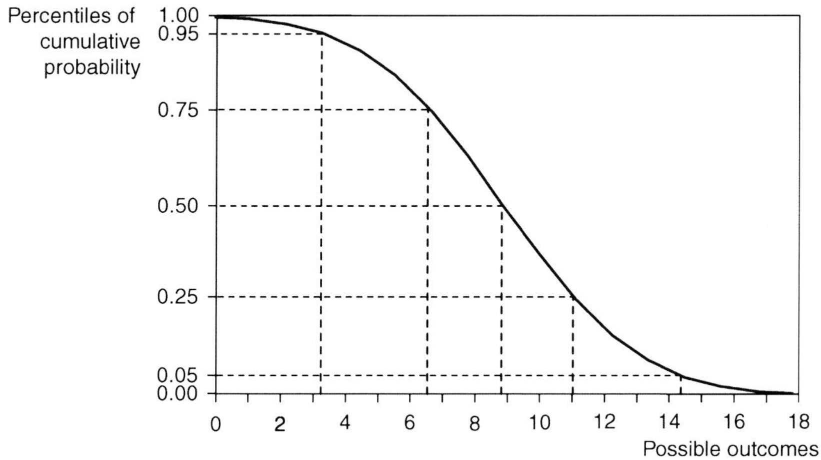

Figure 13.14: A 7-percentile probability distribution (displayed as an inverse cumulative distribution).

75%, 95%, and 100%. The data for a multiple percentiles distribution may, for instance, come from the observation of a previous case assumed to be similar to the case at hand. Figure 13.14 shows a cumulative distribution defined by 7 percentiles, where the outcome is considered certain to be in a range of 0–18.

In literature, references can be found to several other distribution forms which will not be reviewed here. Among them are *beta* and *Weibull* distributions, whose shapes can resemble lognormal distributions but are defined differently, and *exponential* distributions, whose density distributions are strictly falling or rising. There are also several distribution forms which apply specifically to discrete distributions, such as the *binomial* distribution and *Poisson* distribution.

13.6 Summary

The income statement and the balance sheet are the key pieces of information on a firm's financial situation. They are produced for corporations consisting of many affiliates, as well as for the constituent companies. For petroleum firms, the supplementary information on oil and gas reserves, found in the annual report, is also of great importance for assessing the firm's future as an oil and gas producer.

Companies have accounting systems and procedures enabling them to aggregate all transactions into the financial reports as well as other purposes such as cost recovery and managerial control.

Cost estimates are essential when planning a field development project. They are the basis for the profitability analysis which is the basis for accepting or rejecting a project. Analytic estimation has become the more common method, whereas synthetic estimation is an alternative approach that requires less documentation. The costs of a project are divided into Capex, which include investment and drilling costs, and Opex which include the operational costs and transportation tariffs. For estimating the Capex, class requirements and breakdown structures are used. Operating concepts should be optimized with regard to total costs, which can be supported by operating cost estimates which are prepared early and subsequently updated.

The key method of assessing the economic attractiveness of a possible investment is to use cash flow discounting to arrive at a NPV and an IRR. Assumptions and uncertainties must be carefully considered.

Probability distribution is a quantitative approach to handling uncertainty for analytic purposes. Common probability distribution functions include the normal, lognormal, triangular, and uniform distributions.

References

Brigham, E.F. & Ehrhardt, M.C. 2007. *Financial Management: Theory and Practice*, 12th ed. Thomson South-Western, Mason, Ohio, USA.

Laughton, D. 1998. The potential for use of modern asset pricing methods for upstream petroleum project evaluation; introductory remarks. *Energy Journal*, 19(1), pp. 1–11.

Luehrman, T. 1998. Strategy as a portfolio of real options. *Harvard Business Review*, 76(5), pp. 87–99.

Responsibilities to society and business ethics

Commercial firms pursue the goal of making profits. People engaged in the petroleum industry sometimes face the following question: are profits the paramount goal of what we do, or should we also pursue other goals even if it may reduce profits? This question is also posed by the society and by individuals in the society.

Petroleum firms often find themselves in situations where they need to deal with this and related questions. This chapter will expose some of these questions. The complete answers cannot be provided here, nor probably in any book, as these questions need careful consideration by each participant in the industry.

14.1 Profits, market, and society

Most business in the world is conducted in relation to markets. The concept of a market implies that the participants (sellers and buyers) are free to transact in the pursuit of their own profit. Goods are bought and sold whenever a willing buyer and a willing seller agree on price and other terms. Goods will be produced and supplied when they can fetch prices that exceed the costs of production and delivery.

14.1.1 The market and external effects

The Scottish philosopher Adam Smith (1723–1790) introduced the concept of the *invisible hand of the market*. It implies that markets tend to function so as to create a harmonious condition for the benefit of all, even though each participant only pursues his own selfish interest. If this were universally the case, then all creation of wealth might be left to markets and their self-interested participants, with no need for any guidance other than to seek profits and honor contracts.

Markets may be effective for creating wealth, perhaps even harmony, to a large extent, but not universally. First of all, business cannot function well without the rule of law. Further, business activities often have impacts on people other than those who chose to engage in the business. Safety and environmental issues, as discussed in Chapters 10 and 11, are examples of areas in which business activities may have negative impacts on other parties and on society in general. Economists refer to this phenomenon as *market failure*, that is, a misalignment of interest because someone is able to inflict a burden on others without having to pay for it. Legislation and regulatory activities by the state are imposed as a defense against the potential negative external effects that the petroleum industry might otherwise impose on others.

The effects of the petroleum industry on the general society should be predominantly positive. The industry's basic contribution to society is to produce and supply petroleum, without which modern society can hardly function. The extractive, "upstream" part of the petroleum industry utilizes resources which are non-renewable and usually claimed by the nation where they are located, and the nation should therefore receive a large part of the value derived

Figure 14.1: A positive external effect of petroleum activities. To carry out exploration work near the Lake Albert in Uganda, a new road was needed and was built by oil companies. The road made it possible for villagers near the lake to develop a business of exporting fresh fish, thereby creating a new source of income.

from these resources as discussed in Chapter 12. Moreover, the petroleum activity itself often creates new opportunities for gainful employment and increased wealth, and can have other favorable indirect effects as indicated by example in Figure 14.1.

14.1.2 Moral duties and enlightened self-interest

What are then the obligations of firms to society? This issue has been subject to much learned debate. One view defines these obligations narrowly, that business has no other social duty than to increase profits, without committing any deception or fraud (Friedman, 1970). Another view assigns a broader set of obligations to business, including economic, legal, and ethical responsibilities as well as voluntary philanthropic contributions (Carroll, 1979).

Those who subscribe to the first view may pursue petroleum business solely to achieve profits without any regard to society's greater interests, as long as promises are not broken. But oil firms have found that such a strategy may easily be self-defeating. Their long-term success may depend on their business being viewed as acceptable and positive by stakeholders. Stakeholders relative to a business activity are all those parties who directly or indirectly have an interest in the activity and its outcome. There are several examples of this:

- During the first half of the 20th century, Western oil firms obtained large concessions for producing oil in the Middle East and elsewhere. The concessions were increasingly regarded as exploitative and unfavorable to the host countries, and became a source of much resentment against the Western interests. In the 1970s, the host countries, empowered by the rise of OPEC, nationalized the remaining concessions, which seriously threatened the survival of the previously dominant oil firms. Yergin (1991) describes these developments in his

comprehensive account of the history of global oil. Again more recently, national resentment over the terms under which petroleum firms operated has led to upheaval and forced dispossession of assets in countries such as Bolivia.
- Shell's original plans for dumping the Brent SPAR installation to the bottom of the Atlantic Ocean caused widespread displeasure with the public in several European countries. Shell's plan was approved by the UK government, but the company nevertheless felt obliged to change plans in order to avoid damage to its reputation with the firm's customers, employees, and other stakeholders (see also Chapter 9).

As a consequence, most international petroleum firms today make some kind of commitment to pursuing values and goals which are additional to a narrowly defined profit goal. This can be evidenced from statements on their websites and various reports. Following are some areas in which such commitments may be found with petroleum firms:

- protecting human lives and health of the general public;
- protecting the local environment in areas of operation;
- acknowledging responsibility for combating climate change;
- promoting local employment opportunities;
- contributing to education;
- promoting the development of industries of the host country, and creating opportunities for their participation as suppliers;
- promoting general development and social services in the area of operations;
- conducting research and publishing the results;
- preventing the spread of diseases (such as HIV-AIDS);
- commitments to equal rights and opportunities for the genders, ethnic groups, etc.;
- avoiding discrimination of any kind;
- commitment to certain standards of compensation and working conditions for all employees including contractor and subcontractor employees;
- protecting the rights to form labor unions;
- avoiding corruption;
- avoiding investment in countries whose government is illegitimate, has a particularly bad human rights record, or threatens the peace;
- contributions to charities;
- protecting endangered species.

Firms may be under requirements by law or contracts to pursue a number of these commitments. Legislation differs between countries, and many countries have legal systems that allow for practices which would not be tolerated in many other countries. Oil firms tend to operate in many countries. They face the following question: are there certain goals, principles, and standards that we are committed to for all our operations, even if they may be costly and not required by legislation in each country of our operations?

As evidenced by various corporate publications, many petroleum firms indeed express some such commitments. Whether those commitments come from a sense of moral duty or enlightened self-interest is not always clear, and is perhaps not of much practical significance. International petroleum firms must also expect to come under the scrutiny of various activist groups and the press if a large discrepancy should arise between a firm's words and deeds, with the potential for a harmful reputation backlash.

It has become common for Western oil firms to report annually on their achievements in this area. Their websites display, along with the traditional Annual report on financial performance, another report on social issues which may variably be titled Sustainability report, Corporate citizenship report, Environment and Society report, Corporate Social Responsibility (CSR) report, or something similar.

CSR and *corporate citizenship* are related concepts, indicating that a firm accepts commitments to social goals and values which are additional to the goal of making profit. CSR has been much elaborated and debated in management literature. Referring to the debate mentioned earlier, it is a concept to which Archie Carroll but not Milton Friedman would subscribe. Corporate citizenship is probably a less controversial idea, advising that businesses, like any good citizen and neighbor, should display common decency and concern for society in general.

14.2 State and corporate roles

Exploration and production of petroleum are usually conducted in a setting that has important roles for commercial firms and authorities. In principle, the roles should be clear: authorities make the regulations, and the commercial firms produce petroleum while complying with the regulations. There are some practical situations where the responsibilities are less clear.

14.2.1 Regulation and self-regulation

The Norwegian approach to safety regulations was reviewed in Chapter 10, describing a shift from prescriptive and detailed regulations toward a more self-regulatory approach. Similar shifts have occurred also in other countries and in other regulatory areas. Marrewijk (2003) describes a triangular relationship between state, business, and civil society, where states have tended to withdraw somewhat from earlier attempts to regulate society in detail (Figure 14.2).

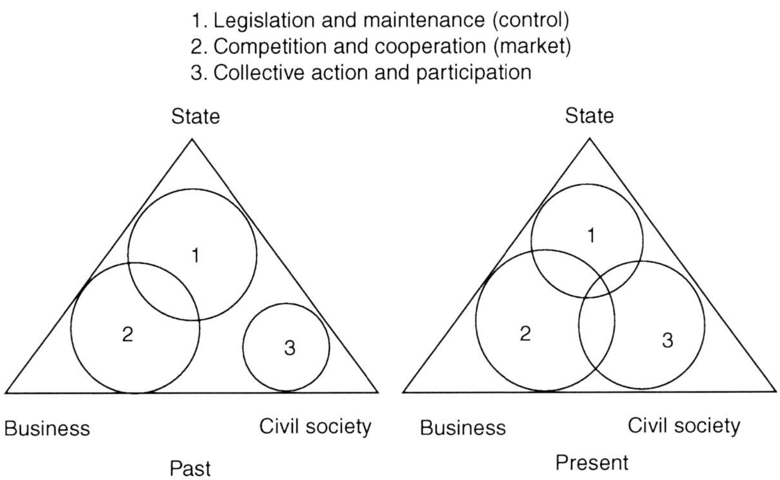

Figure 14.2: State, business, and civil society. Policies and new institutional arrangements support their functioning in areas that were once left to others, redefining their roles and relationships with others (Marrewijk, 2003).

State regulatory agencies may step back from undertaking prescriptive regulations due to the sheer complexity of the business. It is the business of oil firms to design and construct petroleum installations, applying their ingenuity and international expertise, and it may make little sense for a regulator with limited resources to instruct the firm on the details of constructing such installations. Industry firms are generally more at ease with a self-regulatory approach, which imposes less bureaucracy and fewer restrictions on finding the optimal solution.

However, a self-regulatory approach requires a disciplined and constructive approach on part of the firms, and sometimes the willingness to accept a larger cost than a narrow interpretation of regulations might require. In many cases, oil firms have formed industry associations on a national level, as a forum for developing industry quality standards which are adhered to by companies but accepted and monitored by authorities.

14.2.2 Corporate responsibility vs. national sovereign rights

A company's commitment to social goals may sometimes complicate its relationship with the host government of its operations. The firm may be committed to principles and goals that are culturally conditioned from its home location, and at odds with either the culture or political priorities in the host country. A good helping of diplomacy may sometimes be required to handle some potentially delicate issues.

It is normally recognized that a large share of revenues from oil and gas production should accrue to the host nation, in the form of production sharing, taxes, etc., and that the host government is entitled to collect and control such revenues on behalf of the nation (Chapter 12). A dilemma arises if the government uses the revenues in ways which, seen from the outside, appear as unwise or even abusive. A foreign firm may be perceived as having some responsibility for such a situation, and may even look for ways of influencing it for the better.

However, the oil company may in such case effectively reallocate national resources in a way that should be the government's prerogative. Oil firms may be accused of undue interference in politics, something for which there is little tolerance in many countries with a colonial history. Firms must seek a fine balance: they must uphold the commitment to the firm's global values, while at the same time not unduly interfere in political processes of host countries.

14.3 Implementing transparency and avoiding corruption

Corruption and other forms of misappropriation of value are recognized as a serious problem in the world of oil. Estimates have been published relative to certain countries of several billion dollars having ended up in private illegitimate hands that should have been made available for creating lasting benefits for the population. There can be little doubt that a potential for corruption exists in the interaction between commercial firms and state officials where decisions of large economic consequence are being made.

Leading petroleum firms have stated that they are able to conduct business without engaging in corrupt practices, ant that they take a principled approach to the matter. Some countries have legislation making it a criminal offence for their citizens to engage in corrupt practices in other countries, such as the Foreign Corrupt Practices Act of the United States.

The Extractive Industries Transparency Initiative (EITI, website www.eitransparency.org) was launched in 2002 as a joint undertaking of governments, companies in petroleum and mining, and civil society organizations. It is based on the recognition that money from those industries often is associated with poverty, conflict, and corruption. Commonly referred to as the "resource curse," this is often driven by a lack of transparency and accountability around the payments that companies are making to governments. EITI seeks to create that missing transparency and accountability, and has the support of many of the world's largest petroleum companies. EITI has set six criteria for its implementation:

(1) Regular publication of all material oil, gas, and mining payments by companies to governments ("payments") and all material revenues received by governments from oil, gas, and mining companies ("revenues") to a wide audience in a publicly accessible, comprehensive, and comprehensible manner.
(2) Where such audits do not already exist, payments and revenues are the subject of a credible, independent audit, applying international auditing standards.
(3) Payments and revenues are reconciled by a credible, independent administrator, applying international auditing standards and with publication of the administrator's opinion regarding that reconciliation including discrepancies, should any be identified.
(4) This approach is extended to all companies including state-owned enterprises.
(5) Civil society is actively engaged as a participant in the design, monitoring, and evaluation of this process and contributes toward public debate.
(6) A public, financially sustainable work plan for all the above is developed by the host government, with assistance from the international financial institutions where required, including measurable targets, a timetable for implementation, and an assessment of potential capacity constraints.

14.4 Standards of ethics for the engineering profession

Much work concerning the planning and implementation of petroleum project is done by members of the engineering profession. A code of professional ethics has been developed by the Society of Naval Architects, Marine Engineers and Ocean Engineers, recognizing the important consequences of the quality of engineering work. The code was first launched in 1923, last updated in 2000, and can be an appropriate guide also for other branches of the engineering profession. The fundamental principles and the specific canons of this code are exhibited below.

The Code of Ethics for Naval Architects, Marine Engineers and Ocean Engineers (Society of Naval Architects and Marine Engineers)

Fundamental principles
Naval Architects, Marine Engineers and Ocean Engineers or The Society's Professionals maintain and advance the integrity, honor, and dignity of their professions by:

- using their knowledge, experience, and skill for the enhancement of human well-being and as good stewards of the environment;
- striving to increase the competence of the professions of naval architecture and marine engineering;
- being honest and impartial, and serving with fidelity the public, their employers, and clients.

Specific canons

(1) The Society's Professionals shall carry on their professional work in a spirit of fairness to employees and contractors, fidelity to clients and employers, loyalty to their country, and devotion to the high ideals of courtesy and personal honor.
(2) The Society's Professionals shall hold paramount the safety, health, and welfare of the public in the performance of their professional duties. They will interest themselves in the public welfare, in behalf of which they will be ready to apply their special knowledge, skill, and training for the use and benefit of mankind.
(3) The Society's Professionals shall refrain from associating themselves with, or allowing the use of their names by, any enterprise of questionable character.
(4) The Society's Professionals shall advertise only in a dignified manner, being careful to avoid misleading statements.
(5) The Society's Professionals shall regard as confidential any information obtained by them as to the business affairs and technical methods or processes of a client or employer.
(6) The Society's Professionals shall inform a client or employer of any business connections, interests, or affiliations that might influence their judgment or impair the disinterested quality of their services.
(7) The Society's Professionals shall refrain from using any improper or questionable methods of soliciting professional work and will decline to pay or to accept commissions for securing such work.
(8) The Society's Professionals shall accept compensation, financial or otherwise, for a particular service, from one source only, except with the full knowledge and consent of all interested parties.
(9) The Society's Professionals shall build their professional reputations on the merits of their services and shall not compete unfairly with others.
(10) The Society's Professionals shall perform services only in areas of their competence.
(11) The Society's Professionals shall cooperate in advancing the professions of naval architecture and marine engineering by exchanging general information and experience with their fellow naval architects and marine engineers and students, and also by contributing to the work of technical societies, schools of applied science, and the technical press.
(12) The Society's Professionals shall continue their professional development throughout their careers and shall provide opportunities for the professional development of those naval architects and marine engineers under their supervision.

14.5 Summary

Business activities have consequences not only for those who have chosen to engage in them, but also for other people, society in general, and the environment. State legislation and regulations have an important role in obtaining benefits for society and avoiding negative impacts, but so have also the commitments of firms to goals other than making profits.

Many modern petroleum firms make such commitments, and regularly report on their achievements in this regard. One reason for this is their need for good relations with governments and other stakeholders.

The regulation by governments of sophisticated business activities often needs to include a large element of self-regulation rather than relying on detailed prescriptions. This requires a disciplined and constructive role by industry firms, acting sometimes through industry associations.

The SNAME code of ethics may serve as a guide to professional conduct for engineers and other groups working in the oil industry sector.

References

Carroll, A.B. 1979. A three-dimensional conceptual model of corporate performance. *Academy of Management Review*, October, 497–505.

Friedman, M. 1970. The social responsibility of business is to increase its profits. *New York Times Magazine*, September 13.

Marrewijk, M. 2003. Concepts and definitions of CSR and corporate sustainability: between agency and communion. *Journal of Business Ethics*, 44: 95–105, Kluwer Academic Publishers.

Yergin, D. 1991. *The Prize.* Free Press, New York.

Index

WITPRESS ...for scientists by scientists

Materials Characterisation IV

Computational Methods and Experiments

Edited by: ***C.A. BREBBIA****, Wessex Institute of Technology, UK and* ***A.A. MAMMOLI****, The University of New Mexico, USA*

Until recently, engineering materials could be characterised successfully using relatively simple testing procedures. As materials technology advances, interest is growing in materials possessing complex meso-, micro- and nano-structures, which to a large extent determine their physical properties and behaviour. The purposes of materials modelling are many – optimisation, investigation of failure, simulation of production processes, to name a few. Modelling and characterisation are closely intertwined, more so as the complexity of the material increases. Characterisation, in essence, connects the abstract material model with the real-world behaviour of the material in question. Characterisation of complex materials therefore may require a combination of experimental techniques and computation.

This book contains papers from the Fourth International Conference, which brought researchers who use computational methods, those who perform experiments, and of course those who do both, in all areas of materials characterisation, to discuss their recent results and ideas, in order to foster the multidisciplinary approach that has become necessary for the study of complex phenomena. The papers in the book cover the follow topics: Computational Models and Experiments; Mechanical Characterisation and Testing; Materials Characterisation and Testing; New Methods; Advanced Materials; Cements; Porous Construction Materials.

WIT Transactions on Engineering Sciences, Vol 64
ISBN: 978-1-84564-189-4
Published 2009 464pp £176.00

WIT*Press*
Ashurst Lodge, Ashurst, Southampton,
SO40 7AA, UK
Tel: 44 (0) 238 029 3223
Fax: 44 (0) 238 029 2853
E-Mail: witpress@witpress.com

WIT PRESS ...for scientists by scientists

Computational Methods and Experiments in Materials Characterisation III

Edited by: ***A. MAMMOLI****, The University of New Mexico, USA and* ***C.A. BREBBIA****, Wessex Institute of Technology, UK*

As material technology has advanced, interest has grown in materials possessing complex meso-, micro- and nano-structures, which to a large extent determine their physical properties and behaviour. The purposes of materials modelling are many: optimization, investigation of failure, simulation of production processes, to name but a few. Modelling and characterisation are closely intertwined, increasingly so as the complexity of the material increases. Characterisation, in essence, is the connection between the abstract material model and the real-world behaviour of the material in question. Characterisation of complex materials therefore often combines experimental techniques and computation.

This book publishes papers presented at the Third International Conference on Computational Methods and Experiments in Material Characterisation. Topics covered include: Composites; Ceramics; Alloys; Cements and Cement Based Materials; Biomaterials; Thin Films and Coatings; Advanced Materials; Imaging Analysis; Thermal Analysis; New Methods; Surface Chemistry, Nano Indentation; Continuum Methods; Particle Models; Damage Mechanics; Innovative Techniques; Stochastic Methods.

WIT Transactions on Engineering Sciences, Vol 57
ISBN: 978-1-84564-080-4 eISBN: 978-1-84564-279-2
2007 464pp £155.00

WIT eLibrary

Home of the Transactions of the Wessex Institute, the WIT electronic-library provides the international scientific community with immediate and permanent access to individual papers presented at WIT conferences. Visitors to the WIT eLibrary can freely browse and search abstracts of all papers in the collection before progressing to download their full text.

Visit the WIT eLibrary at
http://library.witpress.com

WIT*PRESS* ...*for scientists by scientists*

Electrochemical Process Simulation III

Edited by: ***C.A. BREBBIA*** *and* ***R.A. ADEY****, Wessex Institute of Technology, UK*

Electrochemical processes are used by engineers to protect structures against corrosion, to apply coatings and paints, and as a manufacturing process. However, until recently, engineers had to use experimental testing or frequent surveys to ensure the adequacy of a design as sophisticated prediction models were not available. The papers presented at this conference demonstrate the major advances that have been made in computational modelling to enable the most complex processes to be simulated.

WIT Transactions on Engineering Sciences, Vol 65
ISBN: 978-1-84564-192-4 eISBN: 978-1-84564-369-0
2009 208pp £79.00

Simulation of Electrochemical Processes II

Edited by: ***V.G. De GIORGI****, Naval Research Laboratory, USA,* ***C.A. BREBBIA*** *and* ***R.A. ADEY****, Wessex Institute of Technology, UK*

This book contains papers presented at the Second International Conference in this successful series, which presents and discusses the state of the art of the computer simulation of corrosion, electrochemical processes and the electrical and electromagnetic fields associated with them.

Featured topics include: Cathodic Protection Systems; Modelling Methodologies; Electrodeposition and Electroforming; Modelling of Coatings; Modelling Stress Corrosion, Cracking and Corrosion Fatigue; Modelling and Corrosion of Surface Coatings; Interference and Signature Control; Anodic Protection; Electrocoating and Plating; Optimisation of Control Systems; Detection and Monitoring of Corrosion; Measurement Techniques; Fuel on Photovoltaic Cells; Electrolysis Reactors; Comparison of Experimental Measurements and Computer Results; and Case Studies.

WIT Transactions on Engineering Sciences, Vol 54
ISBN: 978-1-84564-071-2 eISBN: 978-1-84564-272-3
2007 368pp £115.00

WIT PRESS ...for scientists by scientists

Surface Effects and Contact Mechanics IX

Computational Methods and Experiments

Edited by: ***J.T.M. de HOSSON****, University of Groningen, The Netherlands and* ***C.A. BREBBIA****, Wessex Institute of Technology, UK*

The importance of contact mechanics and surface effects in modern engineering and their interaction led the organisers to reconvene this 9th International Conference, which covers two related topics: Computational Methods and Experiments in Contact Mechanics and Surface Effects.

Nowadays the importance of contact and surface problems in many technological fields is well understood. They are complex and inherently non-linear due to the moving boundary and the different properties of materials, particularly along the contact surfaces. Structural components fail by wear, corrosion, high cycle fatigue etc, that is to say they are affected by surface conditions. The use of surface treatments can reduce the cost of components and extend the life of the elements for surfaces undergoing contact, a problem that is specifically addressed by the Conference.

This book contains papers that discuss the following topics: Surface Treatments; Thick Coatings; Thin Coatings; Surface Problems in Contact Mechanics; Indentation and Hardness; Fatigue; Numerical Analysis; Applications and Case Studies.

WIT Transactions on Engineering Sciences, Vol 62
ISBN: 978-1-84564-186-3 eISBN: 978-1-84564-363-8
2009 288pp £109.00

All prices correct at time of going to press but subject to change.
WIT Press books are available through your bookseller or direct from the publisher.